Copyright © 2013 by Action Engineering, LLC / Jennifer Herron
Cover design by Amanda Struz
Edited by Sarah Massey-Warren
Electronic Publication Editing by Travis Heermann
Images licensed from 123rf.com and dreamstime.com

All rights reserved.

No part of this book may be reproduced in any form or by any electronic or mechanical means, including information storage and retrieval systems, without permission in writing from the author. The only exception is by a reviewer, who may quote short excerpts in a review.

Printed in the United States of America

First Printing: December 2013
ISBN-978-1-49487-717-0

Re-Use Your CAD

The Model-Based CAD Handbook

Learn how to create, deliver, and re-use CAD models in compliance with model-based standards.

J.B. Herron, BSME & MSCE
Action Engineering

1st Edition – Electronic Publication

Foreword

After 30 years of CAD data proliferating throughout most industries, Model-Based Engineering requires an Enterprise to re-think their CAD and PLM strategies. In order to optimize the "Reuse" and "Standardization" of CAD, solid models are boiled down to 3 part types: "your parts" (Intellectual Property), "someone else's parts" (Supplier), and "standard parts" (fasteners, fittings, etc).

Traditionally, keeping vaults of part data (be it geometry or meta data) has slowed down organizations, resulting in the proliferation and duplications of parts, thereby increasing costs of the overall product design lifecycle. Evaluating where you have been with lessons learned, and where you want to take the organization with new strategies, requires moving from denial to acceptance of the existence of a CAD problem. Regardless of whether your organization is single-CAD threaded, or multi-CAD enabled, now is the time to learn more about how you can improve your organization in preparation for the next 30 years.

The *Re-Use Your CAD Handbook* provides a foundational understanding of the needs of a model-based infrastructure, including how to centralize and re-use your CAD database. Illustrated with relevant, real-world examples, this handbook offers guidelines to leverage your CAD models to meet today's economic challenges.

— *Tim Thomas, CEO of* CADENAS Part Solutions *(The Standard in Reuse)*

Prologue

Adorning a blue Fiesta dish is a green and blue concoction of fried bananas, over-easy egg, blue corn tortillas, black beans, tomatillo salsa, and *queso fresco*. Over a breakfast of *huevos motuleños*, my friend and I noodle over the similarities between my business in CAD modeling standardization and her business in health-care.

We in America are just beginning to comprehend the enormity of the cost of health and have yet to institutionalize standards for effective use of electronic health records. In the 1980's, Europeans set up a standard for electronic health records that has ensured their safe and efficient use, while also reducing health care cost in Europe.

Electronic health records are an apropos metaphor to explain the tricky, un-standardized business of using digital data to document and protect our Intellectual Property (IP).

So what is it that we Americans are missing about institutionalizing digital product definition (pulling the good stuff from a drawing and adding it into a solid model)?

There are three major barriers: 1) Decision makers need more information necessary to allocate resources to properly implement a model-based environment; 2) Newly released model-based standards are not institutionalized; 3) There are holes in the 3D data process and technology tapestry. The wool is generally available to create the tapestry, but the pattern or the weaver might be missing.

This handbook creates a pattern that illustrates the entirety of the model-based environment tapestry. Protocols are presented for CAD modeling rules and best practices, independent of CAD format. Using straightforward language, strategies are offered to educate the reader on setting-up an effective CAD model-based environment.

Though the technology is vitally important, adoption of a model-based ecosystem requires a culture shift. Clearly, easier said than done, as people are resistant to change. The handbook pays careful attention to culture change needs and promotes good systems engineering practices for model-based implementation.

As *huevos motuleños* layers together simple ingredients to create an explosion of flavor, so too, robust solid CAD models, rich in data, allow engineers to layer together product models, resulting in a complete, complex, inter-related, easily consumable 3-dimensional representation of a product.

— **The CAD Agnostic Author** who offers special thanks to Advanced Dimensional Management (www.advanceddimensionalmanagement.com)

Table of Contents

[1] HOW DO YOU 'RE-USE CAD'? 1
   What Is Model-Based for CAD? 3
   About This Handbook 4
   What is a Model-Based Enterprise (MBE) and Why Do We Need It? 6
   Standards—On Whose Authority? 11
   Will Your Culture Change? 13
   Suppliers 15
   Barriers to MBE Implementation 16

[2] INTRODUCTION TO MODEL-BASED DOCUMENTATION 17
   Terms and Definitions 17
   Model-Based Manufacturing (MBM) 20
   Model-Based Definition (MBD) 23
   Technical Data Package (TDP) 25
   Authoritative Sources 29
   Model-Based Documentation 30
   Action Plans 33
   What Do I Do Next? 36

[3] MBE PERSONNEL INFRASTRUCTURE 38
   Project Team 38
   Roles and Responsibilities 39
   Designers are the Key 46
   Training 48

[4] MBE SOFTWARE INFRASTRUCTURE 50
   Data Management 50
   3D and 2D CAD Management 54
   Product Release Cycle 57
   Libraries and Catalogues 61
   Software Tools 65
   IT Requirements 67

[5] GENERAL MODEL-BASED SCHEMA 71
   The Schema 73
   Terms and Definitions for the Schema 77
   Interpreting the Rules 81
   CAD Rules 82
   CAD Best Practice 94

[6] PART MODEL SCHEMA 103
   Part Modeling Rules 104
   Part Modeling Best Practice 115

Part Template Requirements 120

[7] ASSEMBLY MODEL SCHEMA 121
Terms and Definitions for Assemblies 122
Assembly Modeling Rules.. 125
Assembly Modeling Best Practice 132
Assembly Template Requirements 138

EPILOGUE 140
ABOUT THE AUTHOR... 141

[1] How Do YOU 'Re-Use CAD'?

Today, most product development is no longer a serial, step-by-step process. Product design is generally done by many individuals who work independently yet in parallel with others in a waterfall of design and manufacturing efforts. The trouble is, this method requires an orchestration of both big-picture and minute detail, in order to ensure that all pieces of the final product form, fit, and function together.

Product development collaboration of this order is time consuming and costly; demanding that engineers employ clever methods to achieve a collaborative design process that gains time and does not waste it. A highly effective approach is to leverage 3-Dimensional digital documentation in a structured manner.

We commonly consider CAD (Computer Aided Design) as a means to create part geometry. However, advances in CAD tools give us the ability to leverage 3-Dimensional geometry for more than just 2-Dimensional (2D) CAD drawings. 3-Dimensional data-rich documentation enables both computers and humans to consume product information more effectively.

Though worked on individually, each piece of a 3-Dimensional (3D) product must fit together in the end. Interfaces between the product's pieces are simple to orchestrate when each piece is designed one at a time, one right after the other. Finish the engine shaft, design the gear interface, select a bearing, and design the housing. However, linear creativity is not the reality of product design and certainly not practical for large integrated products such as a car, ship or airplane.

Re-Use Your CAD means we as engineers, technicians, procurement, customers, and more use the same CAD solid model throughout all of our jobs. Therefore, it is critical to use a single definitive data source to ensure data accuracy and consistency as it moves through the lifecycle.

A Model-Based Environment is one where the 3D solid model (parts and assemblies) is the source for: design, analysis, documentation, manufacturing, and much more.

Often full product definition requires additional elements to be appended to the CAD model file to define the entire product information. A documentation method fundamental to that concept is the Technical Data Package (TDP). A TDP is a collection of the CAD model, associated drawing, lists, and required specifications bundled together to convey and archive full product definition. That collection of media is collected in a single "container". The container may be a .zip file or it may be a database item pointing to the source CAD models, drawings, lists, and specifications.

Most engineering problems can be solved with effective communication. 3D model-based documentation is a method to un-ambiguously communicate the product and design intent.

Enrich your organization's product documentation methods by including 3D product definition, rather than relying on flat-to-screen representation of the shape of your part. Two-dimensional orthographic projections remain ambiguous because 2D data is left to a human's interpretation to be rendered into three dimensions. Model-based processes remove the human interpretation, leaving exact data definition. Computer execution and automation occurs using black or white data sets, which are not capable of cognitive interpretation of graphics on a 2D paper drawing.

Use of 3D model definition ensures that the model geometry is completely UN-ambiguously represented. Here lies the catch-22. If all the data necessary to manufacture your product is not sufficiently detailed within the 3D model, it cannot be used downstream to make the part.

We are not losing data; we are finally gaining COMPLETE and UN-AMBIGUOUS definition.

Increased computer assets (greater memory, CPU, hard drive, network, and server resources) allow engineers to expand their capability to utilize 3D data sets for design collaboration. Therefore, it is important to understand exactly what details must exist in the 3D model. As an example, think of getting your model 3D-printed in plastic. The printer prints only the features that are in the digital model. There is no way to point at the corner and say "R.250 inch." If the radius does not exist on the 3D model, it will be printed as a sharp corner.

Careful setup and attention to detail is required to ensure no data is missed as we move from a paper document-based environment into digital three-dimensional data definition. To attempt 3D model-based product definition, without rules in place to govern creation, next assembly usage, data exchange, and use of the 3-Dimensional Technical Data Package, is foolish. This handbook offers model organization, structure and modeling protocols complementing commercial and government standards, to smooth out the model-based tapestry.

What Is Model-Based for CAD?

Model-based means to use 3D CAD solid models to design, analyze, document, and manufacture the design of a hardware product. This method includes not only part models, but also the assemblies these parts comprise.

Ancient shipwrights made models to illustrate their ideas to potential customers. If the model was sufficiently detailed, craftsmen could build a wooden sailing vessel by taking measurements directly from the model and scaling them up. Today we have powerful 3D Modeling software, to assist the designer in explaining difficult concepts and to revise those designs efficiently. However, typically the tools are not used to their full capacity.

If sufficient detail is introduced into the digital model, the software tools can then be employed to produce engineering drawings, and more importantly, accurate 3D models that can be used directly for manufacture. The problem begins with the phrase "if sufficient detail is introduced."

Akin to the issue that we humans lack a protocol to handle the multitude of communication methods bombarding us every day (email, Facebook, Twitter, phone, etc.), we also lack protocols and methods to help us understand and take advantage of CAD software and its tool suite. This book offers that protocol.

Leveraging 3D solid models is not limited to part manufacture. Accurate assembly models can be utilized to provide richer understanding of product assembly instructions.

Model-Based Documentation

The problem? A lack of definition exists for the most efficient way to use 3D models to document a product's design and manufacturing requirements.

Confusion and frustration pervade the area of CAD documentation management and implementation. Many managers cannot understand why the beautiful designs they saw as three-dimensional digital models take so long to come to fruition and are so costly to turn into products. Production engineers ask why they have to laboriously produce 2D drawings of a product that has already been created as a 3D object. Designers wonder why they must constantly rework old models when they could be doing more productive work. To use an overworked phrase, "the Devil is in the Details."

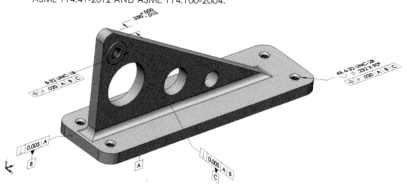

We see in 3D, we should document in 3D.

Currently, existing ASME, MIL-STD, NASA, and ISO standards each address a segment of the MBE system, but not one of these pulls it all together, leaving those designers tasked with implementing Model-Based Engineering a bit stunned.

ABOUT THIS HANDBOOK

Re-Use Your CAD is a model-based CAD-agnostic handbook providing guidance on model-based product definition, the standards that support it, and instructions (independent from CAD software format) for creating part and assembly models that enable Model-Based Engineering.

This handbook will instruct the reader in the following:

- Explanation of the philosophy of designing products using CAD model-based lifecycles

- Implementation guide to model-based commercial (ASME Y14.41) and government (MIL-STD-31000A) standards.

- Model-based benefits, risks, and action plans

- 3D MBD model organizational protocols for part and assembly models, presented CAD agnostically

- Part and assembly modeling best practices

- Setting up, optimizing and maintaining a model-based infrastructure
- An understanding and the benefits of Product Data and Lifecycle Management (PDM and PLM)

The handbook guidelines support the concepts of Model-Based Definition (MBD), as defined in ASME Y14.41, as well as Technical Data Packages (TDPs), as defined in MIL-STD-31000A. Both have a great deal of support in the industry, but have yet to grow legs. This text promotes and illuminates concepts and benefits of MBE through explicit details for CAD templates, metadata, and practical examples. Although PDF drawings as deliverables will be discussed, the primary focus is on Model-Only digital documentation and the methods to achieve a successful delivery.

By providing a 3D CAD modeling standard that supports Model-Based Design, Definition and Enterprise, this handbook will guide experienced and new engineers as well as managers through the maze of technology options, best practices and rules (governed by standards) for CAD models. Protocols presented focus on producing high quality geometry, PMI, and metadata that can be delivered and received accurately.

Having a solid model standard with rules and best practices facilitates a CAD designers' workflow and enables collaboration using solid models among product teams (designers, analysts, manufacturers, quality, and procurement) throughout the product's lifecycle. A structured 3D CAD documentation method frees designers and managers to focus their attention on business value rather than how best to document their designs.

The goal is to shift the engineer's mindset so 3D CAD becomes the source for deliverable documentation and not just a means for creating a 2D drawing. Additionally, engineers can think of pushing the "detailing" of a product into the 3D modeling phase, rather than "detailing" once a 2D drawing is created.

The most efficient way of standardizing the work process is to recognize the 3D CAD model as the source of the design database and the authority from which all other outputs flow (e.g. Finite Element Meshes (FEM), production drawings, bill of materials, and much more). To support this claim, models must be structured for the purpose of coherent documentation and ease of use downstream in the design and production process.

The onus has informally fallen to CAD software developers. However, these software vendors cannot be expected to create the standard because it is not what drives their business model.

Although each CAD system has pitfalls, they all provide value to their customers. This handbook will enhance the user's understanding of CAD systems capabilities (and risks) and provide knowledge for how CAD systems can be implemented to improve their products.

A modeling standard is essential to MBE success. This handbook is intended for use as the foundation for your organization to create your industry- and CAD software-specific model standard through prescriptive model schema rules and best practices.

Who Needs It?

This handbook is rated **E** for Everyone who wants to understand the 3D model-based environment. It serves as a guide to improve CAD modeling techniques in efforts to reduce design lifecycle time.

CAD Design engineers gain greatest benefit from the modeling rules and best practices, and managers and IT personnel will benefit from the straightforward explanations and guidelines to building a model-based environment.

Once a proper model-based environment is prepared, the CAD Designer is able to focus more on the product itself and less on techniques for effective and non-ambiguous 3D documentation.

What is a Model-Based Enterprise (MBE) and Why Do We Need It?

The 3D models are central to your team's network. The centrality of these models is most critical when you are doing large-scale CAD integration, such as designing an airplane, tank, or ship. Because the models are exposed to the entire team, you need a common structure to build your models, so that everyone is able to view and interpret the data. The 3D model organization is called a *schema*. Think of the organizing of your 3D models as the Fourth Dimension (4D) of your model. A schema is simply a template for organizing the digital data.

Model-Based Engineering: An approach to product development, manufacturing, and lifecycle support that uses a digital model to drive all engineering activities.[1]

Model-Based *Engineering* is the process and Model-Based *Enterprise* is the organization that implements it.

Technology exists that supports MBE. What are poorly defined are the processes required to implement MBE. The following chapters will address these processes.

To assist in the transition into MBE, the DOD Engineering Drawing Modeling Working Group (DEDMWG) offers the MBE Capability Index to assist your organization in setting goals to achieve the desired capabilities. Level 0 is drawing-centric and is not considered model-based. Pay careful attention to which level your organization is interested in tackling and take note that for some operations, even though a 3D model may be delivered to manufacturing, the Return on Investment (ROI) for MBE may not be achieved at levels 1 through 4 because the organization is not an "Integrated Enterprise", but rather is a "Disconnected Enterprise."

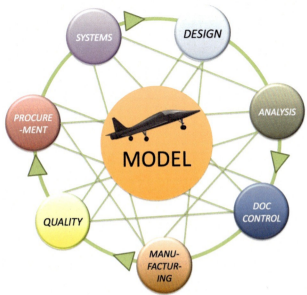

Model-Based Engineering/Enterprise re-uses as much of the model data as possible throughout the lifecycle of the product, being used downstream by many team members and in multiple ways.

"Each level describes a level of capability not maturity. That is to say the levels are focused on defining a set of capabilities for a specific goal. That goal is a business model centered on a set of processes."[2]

Level 0
- Drawing Centric
- Disconnected Manufacturing – Disconnected Enterprise
- Primary Deliverable: 2D Drawing

Level 1
- Model Centric
- Neutral Model CAM – Disconnected Enterprise
- Primary Deliverable: 2D Drawing and Neutral CAD Model

DRAWING CENTRIC

Level 2
- Model Centric
- Native Model CAM – Disconnected Enterprise
- Primary Deliverable: 2D Drawing and Native CAD Model

Level 3
- Model Based Definition
- Native Model CAM – Disconnected Enterprise
- Primary Deliverable: 3D Annotated Model and Lightweight viewable

MODEL CENTRIC

Level 4
- Model Based Definition
- Integrated Manufacturing – Disconnected Enterprise
- Primary Deliverable: 3D Annotated Model and Light Weight viewable

Level 5
- Model Based Enterprise
- Integrated Manufacturing – Integrated Internal Enterprise
- Primary Deliverable: Digital Product Definition Package and TDP

Level 6
- Model Based Enterprise
- Integrated Manufacturing – Integrated Extended Enterprise
- Primary Deliverable: Digital Product Definition Package and TDP via the web

MODEL BASED

MBE Levels as defined by the DOD Engineering Drawing Modeling Working Group (DEDMWG)[3]

Why Do We Need MBE?

In mechanical design, there is no substitute for viewing the data in 3D, because the final product exists in our three dimensional world. In short, you tell the story of your product faster and more completely with 3D data.

There is only one reason for your company to invoke MBE and that is to gain an advantage over your organization's competitors. If you are familiar with the term Value Engineering from the PMPBOK, then you will realize that Model-Based Engineering is equivalent to Value Engineering. Well-implemented Model-Based Engineering methods and protocols save time, reduce risk, and improve products, all of which saves money.

If a picture is worth a thousand words, then a 3D CAD virtual model is worth a trillion!!

There are many practical reasons to use the 3D model directly. Major benefits are listed below and categorized as an engineering or business benefit.

Engineering Benefits[4]

1) **Data Associativity:** Data flows from the 3D model to a drawing or other derivate documentation. Associative connections are maintained among product definition, model, and Next Higher Assembly (NHA).

2) **Automation:** Software can "automagically" process digital data sets and metadata, meaning menial tasks can be automated. For example, software can automatically translate the data set into various derivative neutral formats as needed by suppliers.

3) **Improved Data Exchange:** Because the data sets can "automagically" be translated and verified, less human time is dedicated to manual data exchange. Additionally, automated data exchange results in improved quality of product data set.

Business Benefits[5]

1) **Time Savings:** MBE methods result in time savings of a factor of 3 for first article product development and a factor of 4 for engineering change management (CM).[6]

2) **Data Reuse:** MBE enables effective data reuse across the entire system's lifecycle, improving the efficiency of your organization.

3) **Value of Archived Data:** Because your data is accurate from beginning to end and you have validated the data to achieve a closed

loop solution, your archived data becomes very valuable, instead of "possibly" useful.

4) **Reduced Non-Conformance Cost:** *According to the Project Management Book of Knowledge (PMBOK), there are prevention costs and failure costs. Prevention costs are the cost of building a product that meets requirements. There are also Non-Conformance costs, which result in money spent to fix a defect once in production. Non-Conformance costs are a factor of 10 higher than conformance costs[7]. Attending to reducing errors prior to production saves money.*

Fixing errors while in production costs 10 times the amount of the cost it takes to prevent the errors in the first place.

Realize many more benefits exist than the ones listed here, including benefits to a particular organization. For example, using MBE, you may enable innovation by creating a new product or improving an existing one, making that product much less expensive to manufacture and thus saving you or your organization money.

Because designers are the first people to create a CAD model, they become the core of your MBE process and execution. When designers have tools that are properly configured, they are then allowed more time to collaborate with the team to drive innovation into your products. When management fully embraces and commits to achieving a model-based environment, only then can your organization achieve the efficiencies that MBE can offer.

Risks

No major undertaking comes without risks. MBE has risks, but as always in good system engineering, identifying the realistic risks and putting achievable mitigation plans in place ensures success.

Each organization implementing MBE will discover its own risks based on its business practices. Listed are four common risks.

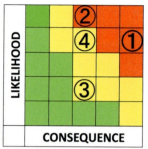

Suggested Probability Impact Diagram (PID) ratings for the four most common MBE risks.

1) High Development and Change Costs

This one is a "biggie". It is very likely to occur and has the potential to create the most damage. Proper or improper planning will make or break your return on investment. Risk mitigation is more complicated because the planning is complicated.

Risk Mitigation: Plan and run MBE Implementation like a regular project.

- Create milestones, action items, and weekly meetings. This is your prevention cost.

- Avoid managing this major process change by submitting an IT (Information Technology) help ticket.

- Evaluate the cost of implementing a new system. The predicted ROI (including risk liabilities) should exceed 66% or higher over 18 months.[8]

- It is critical that transitions happen quickly and effectively to minimize festering bad feelings among users.

2) Data Migration and Interoperability Challenge

Whether your organization wants to migrate a whole project or import a single vendor part into your CAD system (an interoperability challenge), an organized, well-thought out, data interoperability strategy is required.

Depending on whether your MBE process is starting fresh or you intend to pull your legacy data into a model-based system will determine the scale of this risk with regards to data migration, and thus your organization's consequence may rate higher or lower than medium.

However, most likely, you will have some type of interoperability issue, whether it comes from working with another company, delivering data to your customer that is not in your native format, or simply pulling in catalogue parts such as bearings, nuts, rivets, and bolts.

Risk Mitigation: Robust geometry and metadata translation processes, and proper software tools mitigate this risk. This particular risk mitigation strategy must be carefully researched and planned to provide a good match for your company's CAD software suite and business practices. Data translation technologies run the gamut from software that translates multi-CAD format inputs to the multi-CAD format outputs to customized point-to-point translation service solutions. All data translation technologies provide varying quality and quantity of translated data.

3) Limiting Engineering Innovation

If it's possible to make Harry Potter into Darth Vader using a very prescriptive set of interfaces (LEGO® studs and tubes[9]), then setting up interface organization for complicated large systems to fit together should not limit the designer's creativity. In fact, it is quite the inverse. This risk is rated as a low likelihood and a medium consequence.

Risk Mitigation: MBE provides foundational methods and organization to allow users to function together... and 'snap' together their parts. It is the manager's responsibility to inspire users to comply with the standard, so that the parts integrate together as easily as LEGOs® do.

Harry Potter morphs into Darth Potter using only common interfaces.

4) Stalled by Culture Change—A leap of faith is required

At some point, after all the business case analysis is complete, you will have to make a leap of faith and know that a model-based strategy will improve your organization, and the ROI will be met with proper planning. It is difficult to get a warm-fuzzy feeling that MBE is without significant risk, and that is why you must stay on top of the implementation process. There is a high likelihood this risk will occur along with a medium consequence. The risk is created by negative influencers, a culture of "I don't want to. I don't have to. You can't make me."[10]

Risk Mitigation: Mitigating this risk is a management responsibility. Managers must require their employees to comply with MBE goals and standards.

STANDARDS—ON WHOSE AUTHORITY?

Your company's job is to tailor the methods presented by standards, best practice guides, and CAD software capability into processes that your users could execute. For this to occur effectively, management must officially sanction the use of and compliance with chosen standards, guidelines, and possibly the modification or creation of company practices.

> *"If you let everyone proceed in the direction they wanted to, you'd have anarchy."*
> *– Robert Green[11]*

Standards that Govern MBE

Listed below are five areas that group MBE standards over the product lifecycle. Notice some standards have latest releases as early as 2012. As a result, most have yet to be codified by industry, but this handbook will help guide you in using these standards.

3D Data Content and Format
- **ISO 10303-242 (Scheduled Release Jan 2014):** Content and format data exchange using STEP and PLCS
- **ISO 14306: 2012:** Content and format data exchange using JT with Parasolid
- **ISO 14739-1 (Under Development):** Content and format data exchange using 3D PDF with PRC

3D Product Definition
- **ASME Y14.41 (Released 2012):** Digital product definition data practices
- **ISO 16792 (2006):** Digital product definition data practices
- **ASME Y14.100 (Released 2013):** Engineering drawing practices

3D Technical Data Package (TDP Delivery)
- **MIL-STD-31000A (Latest Release 2012)**: Standard practice for Technical Data Packages. How to organize your product data and in what format to deliver it.

3D Model Archival
- **NAS 9300-007**: LOTAR: Long Term Archiving and Retrieval of digital technical product documentation such as 3D CAD and PDM data

3D Model Technical Publications
- **S1000D**: International specification for Technical Publications utilizing a common source database.

> *"If it's easier to be standard than to be nonstandard, people will follow the standard."—Robert Green[12]*

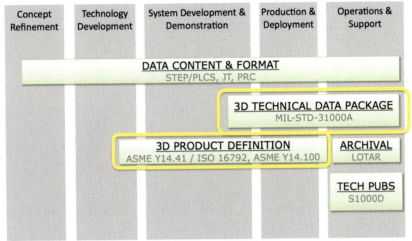

MBE relevant standards mapped against the defense acquisition lifecycle.

Currently the best way to deliver geometry and meta-data is in a standard neutral format. You have a choice of three different neutral format protocols (ISO 10303-242, ISO 14306:12, and ISO 14739-1) that accommodate 3D geometry and PMI (Product Manufacturing Information) data set creation. Those three formats are better known as: STEP, JT, and PRC, respectively. Delivering both native and neutral formats is acceptable, but you must establish a hierarchy for which data set is the governing authority and which is a copy (derivative) of the original data set.

WILL YOUR CULTURE CHANGE?

For the most part, I am a rule follower, which is why it pains me to say that as you follow me on this journey to embrace digital hardware design methods, we must FIRST break the rules in order to re-write the rules.

"Knowing that things could be worse should not stop us from trying to make them better."—Sheryl Sandberg[13]

Consider suspending your focus on CAD software technology and re-focusing efforts on the process and needs of the enterprise. The software is the implementation tool; the process is what your organization creates to matches your business needs.

> *"CAD software is the filter that converts human expertise into digital design files.*
> *-Robert Green* [14]

I will share my experiences that have driven my rule-following personality to break existing traditional rules for product documentation and design to "make them better." When I say 'them,' I mean products. These rules apply to any hardware product where you desire to create GREAT hardware faster, cheaper, and with higher quality.

> *"Talking can transform minds, which can transform behaviors, which can transform institutions."*
> *– Sheryl Sandberg*

The biggest challenge you will face to get a model-based approach adopted is "this is not the way it's done!" Drawings have been used for hundreds of years, so why change? Two reasons: company growth and the ability to attract innovative workers.

Digital product definition is inevitable; you are better off to embrace it now and prepare for the deluge of information that accompanies model-based advantages. To ignore the future of solid model data use is the same as putting your kids in a bubble and not allowing them to watch TV or play an internet computer game. They will do it somewhere anyway. Better to be under your watchful eye and guidance so they don't abuse technology or become endangered by it.

> *"The best way to predict your future is to create it."*
> — *Abraham Lincoln*

My father, S. Russell Herron, wrote a book called *The Danger and the Opportunity*, with the subject of software development in the 1980s. Unfortunately, we are still addressing the same issues today. Better to be prepared for the crazy "Amazing Race" trip, rather than barrel into it unprepared. Today's young engineers will embrace a model-based environment as well as insist on model-based documentation. For your veteran CAD users, you will need a plan to inoculate them with tiny injections of model-based protocols over a longer period of time.

So what is our "Danger" today? If you can't define the manual process, your automated process will fail. You have to tell the computer what to do, as it is not sentient.

Traditionally, the desire to move model-based design processes came from the bottom of the employee food chain. Only some of the information bubbled up to extremely skeptical decision makers. This method has failed to yield a Model-Based Enterprise for many companies. In fact, the only successes are those companies that have driven the change from the top down, demanding that model-based documentation become the priority to the bottom users' daily workflows. If the movement comes from the bottom up, many months, perhaps years will be wasted cobbling together interim solutions. Interim solutions only make your documentation scenarios worse and do not yield any return on investment.

If you are the manger and you don't believe in model-based documentation, your workers' hands will be tied, as there will be no incentive to prioritize MBE as a critical task, meaning their other jobs will overcome MBE implementation and maintenance and your MBE implementation will be in jeopardy.

Throw off outdated conceptions of product documentation and imagine the possibility of a far more innovative, efficient, and competitive method for 3D modeling.

SUPPLIERS

It is common to blame our fear of changing to this new documentation method on claiming that our supplier network (machine shops, material suppliers, etc.) cannot and will not accept 3D data. However, according to the 2012 Supplier Feedback study on the 3D Technical Data Package (TDP), suppliers prefer that you deliver your data to them in 3D. Specifics of elements included in the TDP are covered in Chapter 2.

Feedback on the 3D Technical Data Package (TDP) was gathered from 46 different supplier contacts. The summary of key findings from the study shows that a large majority of suppliers prefer to work directly with the 3D models. The study does not address CAD format or data communication interoperability.

Supplier Demographics:[15]
- 87% are responsible for receiving technical data and models.
- 83% are involved with handling quotes and estimation.
- 80% work with models related to design.

Key Findings:[16]
- 89% of respondents said that the 3D TDP (Technical Data Package) has all of the information that is needed to make a part.

- Most useful features of the 3D TDP, according to respondents are:
    - Imbedded CAD and .STP files (91%)
    - Fully annotated 3D viewable (87%)
- 89% of respondents feel that the 3D TDP is better or much better than 2D drawings for conveying design intent.
- 84% of respondents plan to use the 3D TDP in their manufacturing planning.
- 76% of respondents plan to use the 3D TDP to develop their CAM program.
- 74% of respondents plan to use the 3D TDP as an instrument to convey intent for shop floor.

Because it is your data, I recommend establishing a contractual relationship with your suppliers that identifies specifics of the 3D data transfer such as: permitted translations and interoperability issues. This contract also must address revision updates and a feedback loop for ECR (Engineer Change Requests) that might originate from the supplier against the 3D data sets.

BARRIERS TO MBE IMPLEMENTATION

What are your organization's barriers to MBE implementation? The answer is that it depends. It's possible that your suppliers may be resistant to this change, but according to the NIST Supplier study, they prefer to receive 3D technical information in addition to the standard TDP delivery for product definition.

At this juncture in the evolution of MBE, there are five primary barriers among engineers to adopting MBE and institutionalizing its concepts.

1. Decision makers need more information about Model-Based Engineering technology and implementation techniques in order to effectively seed change within their organization.

2. Holes appear in the technology tapestry of Model-Based Enterprise. It is not obvious to managers and implementers that technology exists to complete the re-use cycle from beginning to end.

3. Standards do not address large-scale assembly integration into MBE.

4. Agreement on authoritative sources needs to be determined. In MBE, CAD models MUST BE the design authority, and that authority starts with the CAD designer.

5. Change is hard.

[2] Introduction to Model-Based Documentation

To coordinate and communicate via a 3-Dimensional design database is the greatest advantage of a model-based approach. Model-based techniques allow organizations to leverage their three-dimensional design database to flow derivative 3D data directly into neutral formats, Finite Element Meshes, drawings, and CNC (Computer Numeric Control) machines, instead of cobbling together an amalgamation of disparate data sets.

First things first... **DO NOT PANIC!**

Shifting from a well-established system of 2D product documentation into another dimension is not trivial. Allow yourself time to absorb the concepts and consider them carefully.

TERMS AND DEFINITIONS

Consistent language is important to use when discussing the subject of MBE. The language is tricky; for instance, a drawing is not a 3D model, and vice versa. What is commonly called a drawing may be either: a 2D representation of the model using orthographic projections that are digitally created using CAD software, or the derivative output which is often a .pdf file or piece of paper.

I can't tell you how many design development meetings I have sat in and heard managers tell their engineers to give them THE drawings. The engineers stare back blankly, calculating the number of hours it will require to put together a drawing with adequate product definition. Somewhere around five business days comes to mind, and the engineer begins to backpedal.

However, what the managers really want is to send the existing 3D Models, available at that very moment, to share the concept with a team member or customer, not fully appreciating the level of maturity of a model in the "concept" phase. This miscommunication drives hours and sometime days of additional work for a single part that wasn't needed into the work schedule. Conversely, managers often do not appreciate the amount of time required to communicate adequate detail to product definition, whether the documentation is 2D or 3D.

This miscommunication drives a wedge into the culture change required in switching to MBE. Understanding the language of 3D documentation and definition is crucial to ensuring that your operations will run smoothly.

Sprinkled throughout this handbook are definitions that describe particular terms, and are often accompanied by an illustration. Further study of MBE terms and their associated definitions (sometimes there are several, gathered from various standards and organizations) can be accomplished at the Action Engineering website (http://www.action-engineering.com/MBEDictionary).

Terms and definitions identified in locations other than in a Terms and Definition section are identified with the following dictionary symbol.

Here are a few terms you should know right away.

Model: A combination of design model, annotation, and attributes that describes a product.[17]

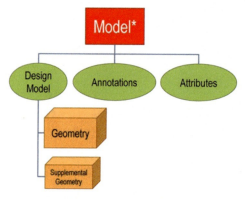

The term Model has many definitions. For this handbook, we will use the term Model as defined in ASME Y14.41.

Annotation: A graphic or semantic text entity that describes the dimension, tolerance or notes of a particular feature in the model. It is stored and visible in the model as an exact and permanent form of digital product definition.

Example of features of size annotations.

Product and Manufacturing Information (PMI): Conveys non-geometric attributes in 3D Computer Aided Design/Manufacturing/Inspection/Engineering (CAD/CAM/CAI/CAE) systems necessary for manufacturing product components or subsystems. PMI may include Geometric Dimensions & Tolerances (GD&T), 3D annotation (text) and dimensions, surface finish, and material specifications. CAx application literature may also refer to PMI synonymously with Geometric Dimensions and Tolerances (GD&T) or Functional Tolerancing and Annotation (FT&A).[18]

Metadata: Data that supports the definition, administrative, or supplemental data package. Metadata includes all relations, parameters, and system information used in a model. This data resides at the model and feature level.[19]

Derivative: Data duplicated or extracted from the original. A copy of a derivative is also a derivative.[20]

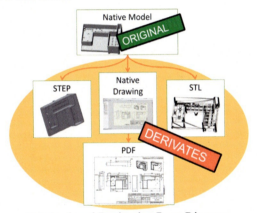

Original and Derivative Data Diagram

Associativity: The established relationship between digital elements.[21]

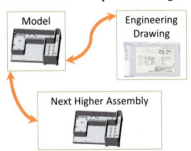

A relationship from one file to another file, or from a feature face to an annotation is an associative relationship.

MODEL-BASED MANUFACTURING (MBM)

Generally, an organization wants to take advantage of sending a 3D model to a vendor to directly manufacture from that delivered 3D model. However, this desire brings the conundrum of sending a 3D model as the manufacturing source of data, while still calling the drawing (usually a digital PDF) the "Design Authority." Existing out in the world are now two files that may or may not conflict, yet define the same product.

ASME Y14.41 demystifies that conundrum by stating that the drawing and model must always agree.

Many engineers mistakenly deliver a 3D model to their manufacturer in order to re-use the model geometry as a starting point for their manufacturing processes. However, they also send a drawing, which they call the product definition authoritative source. If the two data sets, drawing and model, do not agree, then what is the authority?

The ONLY allowable method to make the 2D Drawing your authoritative source is to NOT deliver an associated 3D model.

Authority: Identifies what data serves as the primary and governing source for product documentation. Additional data may be derived from this source, and additional data may be added to the derived data set in order to support the manufacture of the product.

If delivering a 3D model to your supplier is your model-based goal, then all geometry for the part MUST derive from the 3D model. Additionally, the model data and drawing data must be in agreement. Model data trumps the drawing if a conflict arises, because the 2D drawing is a derivative of the 3D model.

Also, dimension data should not be duplicated in the drawing.

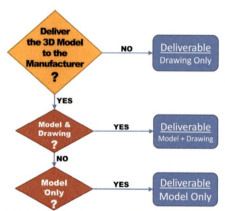

Model-Based Manufacturing Decision Tree
Note: ASME Y14.41 requires the model data and the drawing data to be in agreement.

ASME Y14.41 neither uses nor defines Model-Based Manufacturing, but this is how engineers practically think of it. Engineers and CAD designers have been trained over hundreds of years to deliver a drawing. The development of CAD software arose from the need to digitize that drawing definition. However, CAD technology has now evolved to a place where drawings no longer must be the sole method to define a product in order to have it manufactured. Much more data is available, not only the graspable 3D geometry, but also an overwhelming amount of information about each feature, including surface definition, how that feature should be toleranced, and what material the product is created from. All this data is available digitally, and available for automated processing.

Model-Based Manufacturing (MBM): MBM uses the model created by the Model-Based Definition (MBD) process. It reuses not only the geometric representation of the product contained within the MBD, but also much of the text or "metadata" stored there as well. This reuse eliminates the traditional process of manually recreating this data in order to create the process definition used to produce a product.[22]

Data Set Structure Options

There are four options for documenting product data.

Created by Design Activity	Deliverable	ADV/D/M
1) 2D Drawing *(No 3D Model)* full definition in drawing	Drawing Only	
2) 3D Model + 2D Drawing full definition in drawing	Drawing Only	
3) 3D Model + 2D Drawing partial definition in both; most common	Both	
4) 3D Model *(No 2D Drawing)* full definition in model	Model Only	

Data Set Structure Examples: Courtesy of Advanced Dimensional Management, LLC[23]

1) 2D Drawing

Today the way most organizations document, communicate, and archive their product design is by using a 2D Drawing. The 2D Drawing (generally a digital 2D PDF file) is the record of authority for the product.

2) 3D Model + 2D Drawing, including full definition

This option uses the standard drawing, but a 3D model is used to generate the geometry in the drawing views. Full product definition is only documented in the drawing. The 2D Drawing is still the record of authority. This method is NOT taking advantage of MBE principles, as the engineer must now document in the 3D model and then again in the 2D drawing, manually syncing the two data sets. This method, in fact, is driving more cost into design development and documentation processes. Additionally, there is a greater probability of and duplicate data sets, leading to an increase in data accuracy errors.

3) 3D Model + 2D Drawing

The authority for the 3D Geometry is the 3D model, while the PMI remains on the drawing. This method is a good first step towards MBE implementation because it recognizes the 3D model as the authority for the drawing and separates the PMI to be relayed on the drawing. This structure pulls the attention of the team to focus on the model as the source data for the geometry, rather than purely focusing on the drawing and knowing the 3D model accuracy is now also important. You could also choose to enable 3D PMI in the model and create a derivative copy of the PMI on the drawing.

This practice will prepare data creators to ease into model-only definition. However, the implementation method you choose will depend on your tool sets' ability to create annotations on the model easily and accurately. An additional consideration is the flexibility of your data creators to adopt annotating on a model, rather than on 2D orthographic views.

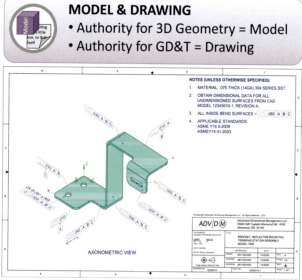

Model + Drawing Example: Courtesy of Advanced Dimensional Management LLC[24]

4) 3D Model

In this data set structure, the authority is the 3D Model and the associated metadata all contained in a single 3D file. Delivered formats may vary from Native CAD (.prt, .sldasm) formats to Neutral CAD formats (STEP, JT, 3D PDF). Using the 3D model as the authority allows organizations to gain the biggest bang for the buck. If it is possible for your organization to jump ahead to model-only, your ROI (Return on Investment) will be realized more quickly.

However, model-only systems come with caution; when done wrong, model-only will be terribly wrong, as data is no longer as concrete as it once was (i.e. on a piece of paper that one can hold). If the humans in your team are not able to review and manipulate the data within their own job's workflow, then attempts to go model-only will fail because the rest of the team will not have access to the data. Orienting team members to the revised presentation of the data is critical. When the data is no longer presented on a piece of paper, our brains (trained for paper documentation) get overwhelmed with where to access the data needed to get the job done.

TIP: When jumping to Data Set Structure Type 4, paying attention to the implementations and practices recommended in this handbook become critical.

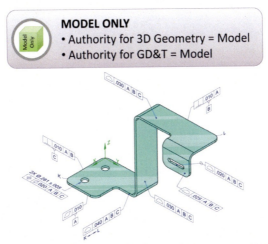

Model-Only Example: Courtesy of Advanced Dimensional Management LLC[25]

MODEL-BASED DEFINITION (MBD)

ASME Y14.41 and MIL-STD-31000A have overlapping requirements that complement each other to help to standardize how to produce model-based product definition.

MIL-STD-31000A defines the entirety of the Technical Data Package (TDP) required to deliver model-only products to the government.

ASME Y14.41 defines what and how geometric dimensions and tolerances should be added and included in the digital product definition of the model.

Most likely, both these standards will continue to evolve over time. This handbook material reflects rules from ASME Y14.41 and MIL-STD-31000A consistent with the date of publish and the status of the standards.

Model-Based Definition (MBD): A 3D annotated model and its associated data elements that fully define the product definition in a manner that can be used effectively by all downstream customers in place of a traditional drawing.[26]

Example of a bracket defined with data structure #4 (model-only) using Model-Based Definition (MBD) techniques.

Product Definition Data (PDD): Denotes the totality of data elements required to completely define a product. Product definition data includes geometry, topology, relationships, tolerances, attributes, and features necessary to completely define a component part or an assembly of parts for the purpose of design, analysis, manufacture, test, and inspection. (See ASME Y14.100.)[27]

Product Definition Data Structure—As defined by ASME Y14.41

The Usability Myth is the belief that digital media simplifies and streamlines jobs. But it is just that, a myth, when the technology is not used properly. Technology implemented without a well thought out, thorough system level framework will cause more chaos and nightmares than having done no improvements at all.

The Product Definition Data Structure, as described by 14.41, is almost identical to the definition of a Technical Data Package. For the purposes of this manual, the term Technical Data Package (TDP) will be used to describe the entire data set. The concept of a Technical Data Package (TDP) provides a framework to digitally organize all the required product data while leveraging 3D documentation methods.

TECHNICAL DATA PACKAGE (TDP)

If your company receives a contract from the government requesting a digital Technical Data Package, you may be a bit overwhelmed by the amount or nature of the requested data. However, requirements for TDP Levels and Types within MIL-STD-31000A are described and selected by the contracting officer from a government organization to choose what type of data the government would like to receive and at what level of detail it should be delivered.

The whole of that data is the Technical Data Package and consists of models, drawings, associated lists, specifications, standards, quality assurance provisions, software documentation, and packaging details.

Technical Data Package (TDP): A technical description of an item adequate for supporting an acquisition, production, engineering, and logistics support (e.g. Engineering Data for Provisioning, Training and Technical Manuals). The description defines the required design configuration or performance requirements, and procedures required to ensure adequacy of item performance. It consists of applicable technical data such as models, drawings, associated lists, specifications, standards, performance requirements, QAP, software documentation, and packaging details.[28]

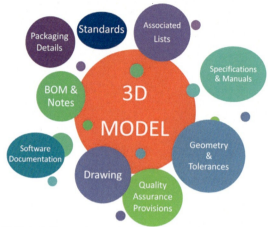

MIL-STD-31000A defines, through a model schema, how to organize those disparate pieces of data into a virtual digital folder.

TDP Type

There are two TDP types, of which this handbook will address only one. The first is 2-Dimensional (2D) Technical Data Package, where the drawing is the product deliverable. The second is 3-Dimensional (3D) Technical Data Package with two subset descriptions: 3D Digital Model-Only and 3D Digital Model plus Associated 2D Drawing. These two subsets will be addressed by this handbook and are consistent with the ASME Y14.41 descriptions of model-only and model plus drawing.

When providing a TDP Type 3D Model-Only data set, an increasing set of requirements is placed on the annotations in the 3D model, and requirements are eliminated in the 2D drawing. The table describes the annotation requirements at each TDP level.

TDP Level

There are three TDP level options. They are: Conceptual, Developmental, and Production. Each TDP level has increasing documentation requirements.

Because we are addressing only 3D Model and 3D Model plus Drawing TDP Types, each TDP level must include an accurate 3D model that contains the full geometric representation of the physical shape of the product.[29] This description is equivalent to the Design Model in ASME Y14.41.

The prime differences between TDP levels are:
- *Conceptual-level documentation* is used for analysis and evaluation of the concept of the product. Hardware beyond breadboard manufacture is not recommended.

- ***Developmental-level documentation*** is used for analysis and prototype fabrication. Hardware manufactured with this level of documentation is intended for test and experimentation only.
- ***Production-level documentation*** is used for procurement and fabrication of the end item.

The annotations needed for each model are to be included in the model and placed in a logical group to assist visibility options of PMI. Ideally, the annotations are associated to the related geometry within the model. Remember an annotation is an explicit dimension or note viewable when PMI is visible in the model.

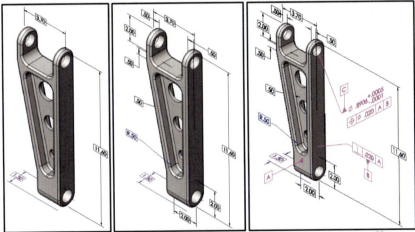

Examples of Conceptual, Developmental, and Production Model[30]

The minimal/conceptual annotated model[31] requires basic overall dimensions and uses block tolerances to define part geometry.

The partial/developmental annotated model[32] requires the annotations from the conceptual model, plus any dimensions that are not covered by the block tolerances and additional critical notes. Also required is a Site Map, which is an index of the available views within the model that the engineer has put together to display all critical product detail.

The full/product annotated model[33] requires the annotations from the conceptual model, plus all dimensions. Profile block tolerances are not accepted. All notes are to be included. Also required are a Site Map and auxiliary views that do not require the user to rotate the model to view the full definition. The product model can be used as the archive.

Annotation	Annotation Description	Conceptual Model	Developmental Model	Production Model
Envelope Dimensions	3 overall boundary dimensions of the part.	✓	✓	✓
Block Tolerances	May be a note that defines all default tolerances to be applied to the product unless otherwise specified.	✓	✓	✓
Material Requirements	Annotation text shall source from centrally controlled material library.	✓	✓	✓
Finish Requirements	May show in notes or displayed as an annotation related to a particular feature surface.	✓	✓	✓
Title Block Information	Number, Description, CAGE Code, Drawn & Approved By, Revision Level and Date*.	✓	✓	✓
Non-Block Tolerance Dimensions	Dimensions shown in an annotation that override model geometry queries. These are required when the block tolerance is not appropriate for a feature. Most commonly used to describe holes.		✓	
Full Dimensions	Defines full product definition.			✓
Site Map	Index of the available views.		✓	✓
Critical Notes	Notes that serve as a note on the developmental model that will turn into the product model full note.		✓	
Full Notes	Defines full product definition.			✓
Auxiliary Views	Provides convenient views to view all required product definition.			✓

*The following items are required data to be included on a title block, either in the model or on the associated drawing.

Annotation Framework Requirements Map for TDP Levels

Product definition data increases as the model moves from conceptual to production TDP level.

Title block data should be set up as metadata variables that can be shared and synced to all data sources (i.e. PDM, PLM, drawing, native solid model, and neutral solid model).

Title Block Data: The following items are required data to be included on a title block, either in the model or on the associated drawing. The most re-usable method of storing title block data is to include it as metadata with the model.

- Number
- Description
- Contract Number
- CAGE Code
- Drawn By
- Approved By
- ITAR Statements
- Distribution Statements
- Classification Requirements
- Revision Level
- Revision Date

On the *TDP Option Selection Worksheet*[34], TDP Types and Levels are selected by the contracting officer to define what detail is required from the models, drawings, and associated lists.

AUTHORITATIVE SOURCES

Documenting with CAD raises these questions:
1. What is the production definition master?
2. What is the Authoritative Source?
3. What is the difference between the two?

We have covered Model-Based Definition and Technical Data Packages, so we know what to deliver, but we do not yet know how to deliver the MBD and TDP as digital product data within the context of our design to release process.

As the design cycle transitions from one phase to the next, the authority for each phase must also transition. The conundrum this creates is how to provide feedback loops to the originating data sets, so the derivative data sets are easily updated.

Providing correct feedback loops requires two authoritative sources, which are interdependent and associative to one-another. Where the product is stationed within the design lifecycle determines which model defines the authoritative source, Design or Manufacturing.

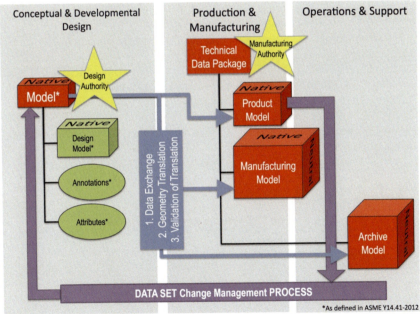

Model Road Map: Identifies the authoritative source for the data as the Model data set moves through the lifecycle?

 Design Authority: The single source authority for the design model, annotations, and attributes.

Manufacturing Authority: The single source authority for manufacturing. It is a synced copy of the Design Authority model. Model data is derived from the design authority, but revised, translated, verified, and synced as required by the change management process back to the design authority.

Product Model: A digital, three-dimensional (3D) representation of an object with precise nominal geometry, attributes, annotations, and fully semantic tolerances that completely and unambiguously define the product and facilitate model-based manufacturing, inspection, and Product Lifecycle Management. The Product Model is the Manufacturing Authority and is a model translated and verified from and against the Design Authority model.

Manufacturing Model: May contain some or all of the Product Model data elements, but will vary based on manufacturing needs of the product. For example, it may be a single 3D solid model in neutral format containing only geometry. The elements contained in the Manufacturing Model are translated and verified from and against the Design Authority model.

Data Set Format

Data set format for the model, product model, manufacturing model, and archive model will be unique to your organization, as your tool set may govern the format required for each stage. However, if you are delivering 3D TDP data per MIL-STD-31000A, most likely the contract will request native and neutral formats. Generally native format selection is at the product developer's discretion. However, neutral format type may be a fixed requirement. The most prevalent neutral format is STEP (ISO 10303), using a container (data wrapper), being 3D PDF.

Note that to date, ISO 10303-AP242 has yet to release.

MODEL-BASED DOCUMENTATION

What do you get for meeting all the challenges of converting from a document-based to model-based product lifecycle? Primary benefits include: increased communication, closed-loop documentation solution, and an organized solution of large systems. These three improvements underlie the goals of aerospace and defense CM (Configuration Management) teams.

Communication

I don't think anyone argues with the importance of communication when developing a large system such as a ship, car, tank, or space station. If it were practical for one designer to build the entire system, then integration challenges would not be paramount, simply because the battles only exist in one person's head.

However, even Orville and Wilbur Wright would struggle with crafting the new Boeing 787 within time and budget constraints. Henry Ford changed automotive history by presenting the world with the assembly line to make the production Model-T fast and cost effective, changing the paradigm of the Ultra-Rich as the exclusive automobile owners.

Overlaying the assembly line scenario onto the design of a battle ship or vacuum cleaner, we take advantage of automating mundane documentation requirement tasks, freeing our engineers to participate in engineering innovation, instead of fussing over documentation processes.

Document in 3D and All in One Place.

Those familiar with the Montessori method for teaching children will know that Maria Montessori created manipulative materials for children to understand math, language, and science, thereby tapping into their brain development.

With model-based documentation, we can extend adults the same luxury that children enjoy, by creating 3D models that are mathematically accurate, rich in detail, and fully define the product hardware. 3D models become a virtual manipulative for designers, managers, manufacturing engineers, users, and maintenance technicians to assist in their comprehension of the hardware product.

When a machinist or assembly technician picks up a drawing, he or she is "learning" the hardware through a virtual medium. Extending the Montessori analogy, a model-based environment provides direct viewing of the geometry, PMI, notes, and metadata in a three dimensional environment, thus stimulating the brain to "learn" the hardware more efficiently and with more clarity. Remember, if the product is documented in 3D and we see in 3D, less translation errors occur even in our human brains if we eliminate having to transpose from 3D to 2D and back to 3D.

Whether your organization's manufacturing and assembly processes are done by real human hands, or your products are assembled via robotic hands, having fewer interpretations from the intended design of the product means the product will be built faster and more correctly.

Hearkening back to the Model-T assembly line analogy, a model-based engineering approach enables all product stakeholders additional tools beyond those traditionally used in 2D drawings to facilitate effective communication of the product design, resulting in fewer errors and less cost spent in "Failure" mode.

A Closed Loop Design System

"I've been signing a lot of online agreements lately." I heard this from a comedian, commenting on the state of our society moving our lives (bank accounts, pictures, Johnny's book report) into databases stored on a server....

somewhere. While I am a huge fan of cloud-based technology (mostly because I love how I can pick up my data from any device and am not chained to a desk to be productive), it does leave a little to be desired in that I don't feel "attached" to my data anymore, and am losing that tactile grip on the data.

The potential to lose touch with the product data scares us because it feels like it is no longer at our fingertips. But just as the cloud allows my photos to be synced to every device I own, data is actually more accessible and more accurate than when it was stored only on one machine.

Cloud accessible data allows us to have, hold, grab, view, and modify in a closed loop back to a single source of data. Bear in mind, I didn't willy-nilly dump my data into the cloud and expect all to be securely synced. I tested a few files, then a few folders, and have now used it long enough to be confident that I am grabbing, viewing, and modifying only the latest data. It behooves organizations also to incrementally implement a model-based approach so as to keep users comfortable with the incremental changes. The goal is to grab, view, and modify model-based data sets with trust that the system operates as expected.

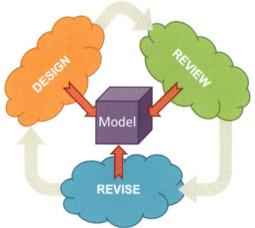

MBE a closed-loop design process for development.

Organized Solution

Many organizations to date have used their Product Lifecycle Management (PLM) software as one of the legs of a stool. Users and workflows comprise the remaining two legs. The product is the stool seat. When the PLM leg is pulled away, so falls your product.

An approach to stabilizing the ever-upgrading and changing software tool change problem is to create a stool where the CAD and PLM software is a fuzzy seat cover for the stool seat that is the product. The three legs remain stable and are: 1) a model-based documentation architecture or schema, 2) users, and 3) workflows.

If an organization relies on the PLM CAD software tools as the foundation to their engineering process, then every time the PLM tool changes, the product quality will suffer.

A well-implemented model-based documentation schema provides an organized environment for which your engineers can innovate within.

ACTION PLANS

Interoperability of data requiring engineers to collaborate properly across design and manufacture workflows is not trivial. It is a complex process, yet one that can be handled by applying scientific techniques.

Create a theorem—Using model-based documentation techniques will save money.

Implement a set of experiments—Start a new product through a model-based documentation cycle or convert an existing product in your Model-Based Enterprise designed system. Your first experiments should start small and progressively increase in size to ferret out issues.

Record and analyze data—Create metrics to help you record important results. Analyze the data results. Interview users and administrators of the MBE system.

Draw conclusions and recommendations—Apply lessons learned from smaller systems to the design of the fully implemented Model-Based Enterprise, then implement the revised process on larger systems and full organizations.

Whether you call the process MBE, MBM, or MBD, there are four required action plans to be successful with a model-based environment.

1) Ensure Accurate Data Sets

Accurate data sets must include geometry, PMI, and metadata. Decide what levels are appropriate: Conceptual, Developmental, or Production. Each 3D model, including assemblies, must be created, checked, and approved to comply with your organization's documented 3D model-based documentation practices.

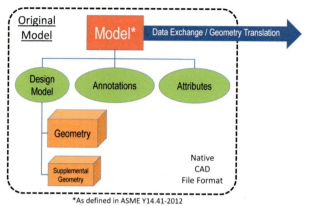

Model Definition for Product Definition Data

2) Enact and Enforce Change Control

All models, parts, and assemblies must be controlled, both original and derived. Every file, at a minimum, must go through a basic data management workflow. In PMBOK or ISO90001 language, you must Plan, Execute, Monitor, and Control all your data, including the 3D data.

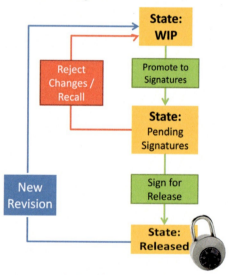

At a minimum, every CAD file must go through this workflow.

3) Define Robust Interoperability Strategies

A reality of a business that uses CAD software is the need to send and receive multiple CAD software formats. It is vital that you not stick your head in the sand on this issue, because as stated in Chapter 1, the likelihood of interoperability challenges is extremely high.

Define a strategy, whether human in the loop or automated, to handle the translations to and from multiple CAD formats.

CAD-to-CAD interoperability provides a significant technical challenge, but layered on top is aligning metadata from the CAD system into the various data management systems and/or CAD-neutral file formats, such as 3D PDF or STEP. For instance, it may seem simple, but maintaining a single part number from the engineering data set, through change management, through quality control, and through ERP systems still poses a significant challenge when data travels through multiple software packages and in varied formats.

Interoperability Clouds

Depending on the organization, any system shown on the diagram may be the elephant in the room for political or cost reasons. Most likely, your organization has already made a significant investment in an Enterprise Requirements Planning (ERP) or Product Lifecycle Management software infrastructure. At the heart of ERP software is a database of YOUR data.

Depending on your situation, you may or may not know or have the capability to modify the user interfaces to access your data, manually or by automation with the ERP. But this is YOUR data, and if the software selected does not allow access to YOUR data, get one that does because you must be able to use YOUR data.

To assist you in creating an interoperability strategy, your organization will need to know the answers to the following questions.

1. With whom and how do you need to interface externally? The answer to this could be external to your design group or external to your company.

2. What is the release process for translated data?

3. Where does the translated data source from?

4. What processes need software tools other than your standard CAD software suite? For example, what tool will you use to validate your translated

data, and how will that translated data be accessed? Do you need a secure web portal to directly access released data?

5. What 3D format is the best for your organization to use in translating data? Do you need to translate multiple formats? What viewer tool is available to view that 3D data?

4) Enable Creative Freedom

Nobody said to Mozart: "How will you document your symphonies? You figure that out at the same time that the music is pouring from your head." Believe me, this is how we designers feel. However, most of us are mere mortals compared with Mozart, and we require a structured documentation process so our brains can focus on the creating.

Provide the appropriate framework and tools, and free your engineers to think up groundbreaking technology.

Straightforward interfaces make the easy stuff easy and leaves time for the hard stuff.

WHAT DO I DO NEXT?

While reading a science fiction novel containing elegant descriptions of the technological marvels of the year 3001, I realized that in all the pages of sci-fi I have read over the years, I can't recall an author who broached the subject of how the scientific marvels were financed. Who paid for them? Fortunately the creative author isn't burdened with the need for economic justification. Do you ever look at the spaceships and walkers in Star Wars and think "Wow that's a really complicated mechanism. I wonder how that was developed and paid for."

The creative design engineer may well be exercising the same neural pathways as the fiction author, but he/she does not have the luxury of ignoring the economics of the process.

Practical budgets will always override utopian ideals. So it is fine to take baby steps when implementing MBE, as tackling MBE often feels like an insurmountable goal, but an increase in model data set re-use is better than very little or no model re-use. Time spent re-creating data that already exists is a waste. Time saved by re-using existing data sets while gaining data accuracy is a major bonus. Keep in mind the solution doesn't have to be perfect—just better.

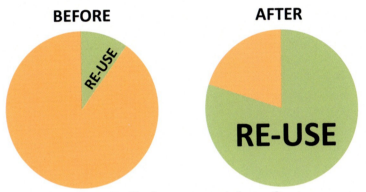

Any accurate Re-Use improvement is better than none.

[3] MBE Personnel Infrastructure

Typically, model-based activities are relegated to a very good CAD designer, who may or may not benefit from the perspective of fluency in multiple CAD software packages. Generally, this assigned super-user will attempt to establish some rule of law among his or her fellow CAD designers, but fall short because that designer has other duties within the company and often has little authority over the other users' priorities.

3D CAD power- or super-users willing to mentor others are very difficult to find. Often, decision-makers don't know what skills, tasks, or responsibilities to request. This chapter serves as a guide to roles and responsibilities required for effective MBE implementation.

PROJECT TEAM

There is no "I" in "team," right? MBE implementation requires a team working in close daily interaction to achieve success. This team is called the MBE Implementation Team.

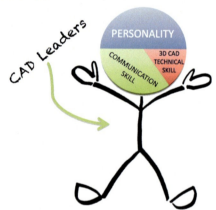

This is not a job for your average engineer.

Typically, a company has two types of users of CAD software: those who perform basic CAD functions, and those who are capable of advanced functions within the CAD software. The advanced users become the "go to" people for green CAD users. By default, the advanced CAD users become the CAD Leaders as they typically have self-starter personalities and LIKE model-based CAD methods.

DO NOT stick someone with the job of model-based implementation who cannot see the benefits of MBE—that strategy will most surely backfire.

Pay CAD leaders competitively, as this particular cost of prevention will be well worth it in money saved from failure cost downstream. Remember, a poor implementation of MBE is worse than no implementation at all.

It is not possible to successfully implement a model-based environment without a dedicated implementation team. The following organization chart illustrates the team needed for MBE implementation success. Discussion below describes tasks combined with skills and responsibilities for each role identified in the organization chart.

Note that each box of the organization chart may or may not represent one FTE (Full Time Equivalent). The FTE allocations are contingent on the number of products and users in your organization. The MBE Implementation Team supports CAD users and team members at a level of effort appropriate to business practices within a particular organization. However, I caution against being stingy with the FTS allocations. If the task list looks too big for one person to cover, it most definitely is.

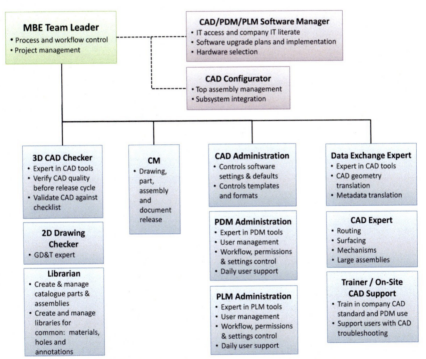

MBE Implementation Team organization chart.

ROLES AND RESPONSIBILITIES

Try as you might to plan, you will never be able to plan for every event that each team member might have to address. Allocate each team member role with additional unscheduled time to cover unaccounted for events. Also, do not forget to include these personnel when calculating the ROI for your organization's MBE business case.

CAD Super-user: A CAD designer with at least 3 years experience modeling 3D parts and integrating assemblies within your entity's particular CAD software tools. Additional CAD software experience, such as large assembly best practice and inter-part relationship understanding, is required to be considered a super-user.

Set up the right implementation team and don't forget the team leader.

MBE Team Leader: We all know good leadership is critical, but it is easy to forget. Although the goal of changing product delivery methods into a 3D environment seems straightforward, the implementation of change is tricky business. A patient, yet energetic soul, willing to use his or her charisma to push your business entity's MBE goals is essential. Successful MBE implementation is not only about good software tools, it is mostly about how the new process is rolled out. In conjunction with charisma, the MBE Team Leader must have the necessary authority to implement the MBE plan. Authority requirements will vary in each organization, but in general the MBE Team Leader will need a budget and an agreed upon method of influence to persuade or motivate those people in your work environment that are outside of his/her direct sphere of influence (MBE Implementation Team). In other words, the MBE Team Leader needs a stick, and the bigger the better.

SKILLS	RESPONSIBILITIES
• Project management	• Hold weekly meetings and track progress
• Communication	• Workflow approval, management and final say
• Budget estimation and tracking	• Liaison to corporate and project management

MBE Team Leader Skills and Responsibilities

CAD/PDM/PLM Software Manager: Most organizations locate their software managers within their IT departments; the software managers are not dedicated to any particular project. Regardless of whether this person's boss resides in the IT department or is the MBE Team Leader, it is critical that the software manager is plugged into day-to-day MBE implementation operations. Although the software is still the fuzzy seat cover to the stool, it has many hooks that must be attached to the seat. Prevent painful user angst by minimizing hiccups when installing or upgrading CAD and PLM software tools.

SKILLS	RESPONSIBILITIES
• Organization IT literate	• MBE implementation IT liaison
• Computer and network hardware knowledge	• Software upgrade planning and implementation
• CAD/PDM/PLM software knowledge	• Hardware selection of CAD workstations
• Database administration skills: SQL, Oracle, etc.	• CAD, PDM, PLM software selection input

CAD/PDM/PLM Software Manager Skills and Responsibilities

CAD Configurator: As the CAD/PDM/PLM Software Manager may not report directly to the MBE Team Leader, so then the CAD Configurator may report directly to a program or project. As the CAD Configurator's primary loyalty will be to the product he or she is designing, this role will normally fall outside the MBE Team Leader's influence. However, it is paramount that CAD Configurators are plugged into day-to-day implementation and maintenance of MBE because they will act as liaisons from the MBE Implementation Team to the CAD designers.

SKILLS	RESPONSIBILITIES
• CAD super-user	• CAD top assembly management
• Systems engineering	• Subsystem form, fit, and function integration
• Design engineering	• MBE Implementation Team interface to CAD users

CAD Configurator Skills and Responsibilities

The organization chart found below has four columns: Standard Enforcers, Change Management, Software Administration, and CAD Experts. In a small organization, a single person could cover all the responsibilities in the respective column or silo. However, bear in mind that the more skills you try to find in a single human being, the more difficult it will be to fill that role. Remember that it may be best to let an individual be good at what he or she does best, and avoid forcing people into accepting responsibilities they do not do well. As an example, expecting a change management processor to become a 3D Drawing Checker may never work, but transitioning a very good traditional GD&T 2D Drawing checker into being a 3D CAD checker may work. Regardless of your company's choice, adequate 3D training will be necessary. Be realistic about employees' time commitments and allow time for them to come up to speed on new tasks.

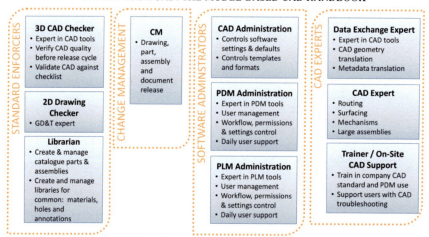

MBE Implementation Team Role Silos

STANDARD ENFORCERS

These are lovingly referred to as the 'CAD Cops'. Within Enforcers, there are three roles: 3D CAD Checker, 2D CAD Checker, and Catalogue Librarian. Depending on the MBE competency of the CAD users, it is feasible these three roles can be accomplished by a single person.

3D CAD Checker: Acting as a mentor to guide users into MBE, the 3D CAD Checker will verify that 3D model parts and assemblies have the correct metadata, PMI, notes and dimensions, and have correct tolerances according to the company or agency standard. In many cases, organizations will choose to invoke ASME Y14.41 and provide additional standard clarification as needed by the business entity. Additionally, the 3D checker will validate that the model was created using modeling best practices and that it integrates into mating parts and next higher assemblies within form, fit, and function guidelines (usually set within a business context). For assemblies, it is important for the 3D CAD Checker to validate that the assembly is built with appropriate mates, inter-part relationships, and has no rebuild errors. Details of recommended part and assembly best practice will be discussed in subsequent chapters.

Ideally, a software tool would automate most 3D and even 2D checker work. However, until checking requirements are fully understood via a human, then you are not ready to automate.

SKILLS	RESPONSIBILITIES
• CAD super-user	• Validate that files comply with part, assembly, drawing, and metadata company/program standards
• CAD modeling best practice mentor	• Validate that modeling best practices were followed
• Design & integration experience— Minimum 3 Years	• Validate that parts/sub-assemblies are properly integrated into Next Higher Assemblies (NHA)

3D CAD Checker Skills and Responsibilities

2D Drawing Checker: If your organization has decided to establish Type 3 TDP, or a Model + Drawing system, then a 2D Drawing Checker is required. Because the drawing details required in MBE architecture are less than a fully defined 2D drawing, the 2D checker job will be less time-consuming than a traditional drawing checker role from the past.

SKILLS	RESPONSIBILITIES
• 2D drawing software expert	• Validate 2D drawings against drawing company/program standards
• GD&T expert	• Validate 2D drawing and 3D model are in agreement.

2D Drawing Checker Skills and Responsibilities

Librarian: A librarian is responsible for producing, archiving, and maintaining any part or subassembly, which is utilized in more than one product within the organization. For instance, if your company builds pan and tilt heads, motors and gearboxes purchased again and again from the same supplier will be catalogue parts. Similarly, common or "standard" parts, such as washer, nuts and screws, are also called catalogue parts. Most likely these parts and subassemblies will be integrated into several of the pan and tilt products. A person who has been, or is, a 3D CAD Checker, easily fills this role. It is advantageous for this person to be experienced with the nuances of CAD software configuration and display options in order to efficiently store common parts.

SKILLS	RESPONSIBILITIES
• CAD software super-user	• Create, import, and clean-up parts and subassemblies
• Attention to detail	• Submit catalogue models for approval from various designers as necessary
• CAD family and design table creation skills	• Manage and maintain catalogue model libraries

Librarian Skills and Responsibilities

CHANGE MANAGEMENT

The change management silo is responsible for tracking and approving revision history of the product. Generally, an existing member of the company already fills this role. The challenge will be to acclimate change management implementers into using new 3D tools to visualize the product data. Therefore, this person must be educated in the chosen 3D visualization tools.

In general, this role remains unchanged during the MBE transition. While the tool change may take time, MBE doesn't necessarily require that the organization upend its change management process. Existing product release processes can still be used but may be implemented using different software tools in order to integrate the 3D data.

Because this role is unique to a business entity, skills and responsibilities are not described.

SOFTWARE ADMINISTRATORS

It is possible to consider CAD, PDM and PLM software administrators as traditional IT roles; however, it is important to keep in mind how complex CAD databases (CAD/PLM Software Tools) are and to allow adequate resources for administrators to provide proper software management. Expertise in each software package is required to appropriately manage these software systems. Using software vendor experts to assist, set-up, and rollout the software is appropriate, but on-site personnel are required to maintain day-to-day operations of the databases. Eliminating in-house maintenance expertise would be like running a power plant without caretakers.

CAD Administration: Significant to the success of internal CAD interoperability within an organization is the consistency of CAD software settings across all user workstations. The task is greatest when software setup or upgrades occur, but will be necessary over time because users, who continuously adjust settings, drive inconsistency into CAD data sets.

SKILLS	RESPONSIBILITIES
• CAD super-user	• Create and maintain CAD software settings and preferences for organization
• Some PDM/PLM administration experience	• Create and maintain CAD templates

CAD Administration Skills and Responsibilities

PDM Administration: Your organization's PDM (Product Data Management), may be the same software tool as the PLM (Product Lifecycle Management) but it is useful to view the product lifecycle workflow through the PDM and PLM separately. Separating these two techniques, PDM and PLM, will help the implementation team in keeping jobs and workflows organized into task categories, resulting in simplifying the confusion between tasks intended for PDM or PLM. Further definitions of PDM and PLM are discussed in Chapter 4.

SKILLS	RESPONSIBILITIES
• PDM administrator trained	• For PDM software, create, implement and maintain: workflows, settings and preferences for network and client software
• Some CAD administration and PLM experience	• User creation and management

PDM Administration Skills and Responsibilities

PLM Administration: Similar skills and responsibilities for PDM administration are needed for the PLM administrator because the PLM is also a database of documents with which the administrator must be familiar. Another reason to view PDM and PLM separately is because your organization's business process may have chosen not to include 3D model data in the lifecycle database. Notice that if it is the case, it will be extremely difficult to keep the data synced and to maintain full lifecycle control of the product when in Model+Drawing or Model-Only digital product definition.

SKILLS	RESPONSIBILITIES
• PLM administrator trained and some PDM administration experience	• For PLM software, create, implement and maintain: workflows, settings and preferences for network and client software
• Change management experience	• User creation and management

PLM Administration Skills and Responsibilities

CAD EXPERTS

The last silo consists of CAD experts who are CAD super-users capable of teaching, mentoring and supporting all users. This silo may be satisfied by 1-3 people, or in small businesses, it may be feasible to spread these tasks over CAD designers tasked with other duties. However, caution is advised when assigning these roles, as careful attention must be paid to the skill sets and responsibilities required. Time commitments for each responsibility may be a challenge to predict and close monitoring and tweaking will be necessary as an organization grows.

Data Exchange Expert: Accurate 3D data set translation is essential to model-based success, as it is unlikely that all data sets will come to your organization in good working order and be fully inter-operable with your CAD software tools. The amount of 3D data sets you receive will dictate the time commitment that should be allocated to the data exchange expert role. Additionally, using effective data exchange protocols between your organization and those delivering or receiving 3D data sets to it may mitigate data exchange time commitments.

SKILLS	RESPONSIBILITIES
• Expertly trained in CAD translation and validation tools	• CAD geometry and metadata translation

Data Exchange Experts Skills and Responsibilities

CAD Expert: Your business may build products that require experts in the areas of surfacing, electrical or piping routing, flat patterns, large assemblies, and assemblies that are mechanisms. If you build more than rectangular boxes, some of these special modeling skills may be necessary. Access to software vendor training for these particular CAD specialties may be sufficient to meet the needs of your designers. If not, employ an on-site or on-call CAD expert in these areas as needed. The need for these experts may come and go as products move from development to production.

The skills required for the CAD Expert role will be specific to the CAD expertise needed. The CAD Expert role will be to assist, mentor, and teach the CAD users in their particular expertise. Your CAD software vendor or VAR (Value Added Reseller) can usually provide an on-site expert.

SKILLS	RESPONSIBILITIES
• CAD super-user	• Unique modeling expertise

CAD Expert Skills and Responsibilities

Trainer / On-Site CAD Support: As is the case with the CAD Expert, other users may need daily help to get up to speed in an area of CAD modeling expertise, or users may be new to a particular CAD software. In this case, having on-site or, at a minimum, on-call technical support is paramount. This expert should be steeped in the company business processes and standards and should hold weekly user meetings to support business processes as applied within CAD/PDM/PLM software systems. Training and communication of regular tips as software improvements or fixes are made. This person is also a good choice to collect service request concerns against the various CAD/PDM/PLM tools and liaise with the software vendor.

SKILLS	RESPONSIBILITIES
• Communication	• Support users day-to-day
• Capable of creating training materials	• Hold weekly user meetings
• Trained in company standards and processes	• Provide CAD training

Trainer/On-Site CAD Support Skills and Responsibilities

At management's discretion is the ability to mix and match skill sets and responsibilities of different roles. The key point of discussing the project team is that each responsibility laid out in this chapter is recognized and accomplished.

DESIGNERS ARE THE KEY

The traditional CAD designer role is to: design hardware, use CAD software to best model the hardware parts and dump his or her brains into the computer in order to explain the hardware designed in his or her head (design intent). In addition to these responsibilities, designers are also expected to be experts in CAD software, Operating Systems (OS), computer hardware used to run the CAD software, the Drafter, the Checker, the CAD data translator expert, and the model maintenance man. Overwhelmed by these additional tasks, designers are often not free to innovate the best solution when presented with a problem.

Establishing a MBE Implementation Team relieves CAD Designers of CAD/PDM/PLM administrative functions, creating and policing company standards and software training. Thus, the MBE Implementation Team allows Designers to think beyond the documentation and concentrate on the engineering.

SKILLS	RESPONSIBILITIES
• Modeling expertise in CAD software germane to their engineering function	• Create models that represent design intent
• Engineering disciplines as necessary	• Follow CAD modeling standards and procedures for parts and assemblies
• Trained in company document release processes	• Use catalogue libraries
• Trained in CAD/PDM/PLM user functions	• Maintain models as changes are incorporated

CAD Designer Role revised with the assistance of a MBE Implementation Team

Designers passionate about model-based engineering will be more efficient, taking pride in their 3D products as well as the hardware product. The basic building block in your design process is the guy with the ideas. As authors are able to use Word Processing software, the designer needs a CAD tool with which to communicate those fabulous ideas.

The CAD software is simply a tool an engineer uses to illustrate what is floating in his head. In order for creative solutions to flow from the engineer's head to be represented in a three dimensional environment, his organization must support his job with an effective model-based work environment.

One of many advantages of a model-based environment is that designers can utilize 3D models to communicate their designs even more comprehensively. Thus, we circle back to the importance of training CAD designers and all MBE users in order to use the 3D model to communicate. This is a major culture shift in design engineering, but one intended to free cycles for creativity, not stunt it.

TRAINING

All the previously mentioned positions, desired skill sets and responsibilities may be useless if all users and stakeholders who touch the data created in a model-based environment must be properly trained.

Training includes not only training in job specific software such as CAD modeling software or CAD visualization software, but also in the defined product lifecycle workflows from model creation to release to archive of the product data.

What do CAD users do with 3D models?
- **Use** CAD software to create 3D geometry
- **Plug in** data as required by company standards and processes
- **Participate** in the process of product creation through release

The MBE Implementation Team becomes the ambassadors of MBE implementation, who create a stable architecture within which users may be productive. Remaining chapters will further describe what constitutes a stable MBE architecture.

Within the product's lifecycle, there are other users who will not be CAD trained. What about these users? What CAD skills will be required of designers and non-designers, and what training will be required to effectively allow them access to the released 3D data sets?

The entire team must have the right skills (training most likely required) on each software tool set used throughout the lifecycle. A plan that only provides basic training to your CAD users and none to other downstream users is ill advised. Moreover, it is imprudent to rely on self-taught designers and downstream users to "figure out" how to use the tools, especially when they do this diddling around while you are paying them. This strategy ensures discordant process implementation and a chaotic 3D model release to production lifecycle.

Users who produce and consume 3D data must know how to do so against the agreed upon company standard.

What do non-CAD users do with 3D models?
- **View** the 3D model data set
- **Plug in** data as required by company standards
- **Participate** in the process of product release cycle

CAD User Certification Program

In order to appropriately train CAD users, a CAD Certification Training Program is recommended. Three levels (Basic, Skilled, and Master) will assist managers when allocating CAD users to a particular task. The skills required to achieve each level are in the organization's specific CAD software package. If multiple CAD modeling software packages are employed, then a level-per-software package is required.

Level 1: Basic
- Basic 2D part drafting
- Basic 3D part and assembly creation

Level 2: Skilled
- Skilled in complex part 2D drafting
- Skilled in part and assembly constraints, organization, and optimization
- Has unique skill sets such as surface or routing expertise

Level 3: Master
- Expert in complex part 2D drafting
- Expert in large assembly 2D drafting
- Expert in complex part 2D drafting
- Expert in large assembly constraints, organization and optimization
- Has several unique skill sets related to CAD modeling
- CAD software administrative experience

Culture Change

A bit of good news about culture change:

"Those who are entering the workforce today are comfortable with change."—John G. Falcioni[35]

In addition to the super-users, configurators, and 3D CAD checkers, projects will use new, less expensive labor to fill routine CAD design work roles. The good news is that fresh-out engineers meet the culture change challenge, reducing the risk identified in Chapter 2 concerning resistance to using 3D models. New graduates will not have issues with using only models to document their designs.

[4] MBE Software Infrastructure

It is easy to think that choosing the right CAD modeling software tool can launch a model-based environment. While the CAD creation tool is certainly important, there are several other large slices, both software and process, required to complete the MBE infrastructure pie.

DATA MANAGEMENT

Data for model-based product design is the information required to produce a particular product. In a CAD model-based environment, product data may be as simple as a single 3D model file in STEP format, or as complex as a top assembly that defines an entire helicopter. Wherever your data falls along the simple to complex spectrum, if it is used to define your product, the most prudent practice is to manage that model data.

Data management is the silver wrapper around the Hershey's™ Kiss. You don't eat the foil wrapper, but there is great stuff inside. The wrapper protects the chocolate from getting dirty until it is ready to reach its final destination — your mouth.

Eating a Hershey's™ Kiss is the same way we consume product data. We don't consume the wrapper, but it is still required to deliver the product safely and cleanly to the proper end user.

There are two major uses for that silver wrapper (data management) along the product lifecycle.

1) ***Collaborate*** *with internal or external team members, customers, and vendors. 3D data collaboration usually occurs before the product is formally released.*

2) ***Manage data*** *(product documentation) that is formally released 3D data sets in order to safely deliver to manufactures and customers.*

Because 3D data is inherently rich in information, it is prudent to track and manage this data in a database capable of complete 3D data set tracking. This is typically done in a Product Data Management (PDM) tool.

Product Data Management: The tracking, control, and status control of product data, either as files or through a database. The usage is frequently confined to the design stage of product development, although it may extend to planning for manufacture.[36]

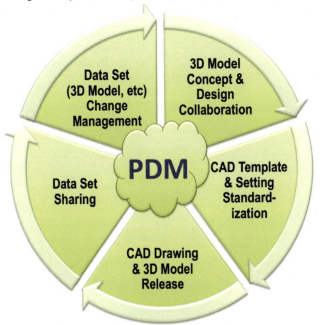

A PDM is the first step to a PLM. Given current technology, PDM systems that complement a native CAD software system are the most accurate and efficient method to control 3D model data.

A PDM designed to manage CAD data can effectively be utilized to manage design collaboration with 3D data sets (Data Management Goal #1) between internal and external team members. Appropriate security protocols are required for internal and external team members to access the source data securely. This usually means setting up limited web portal access to the PDM database.

 Product Lifecycle Management: The tracking, control, and status control of product data, either as files or through a database. PLM extends the scope of PDM to include each physical product produced from a design, including deviations from the design and usage of the product.[37]

Some software suites combine PDM and PLM functions together.

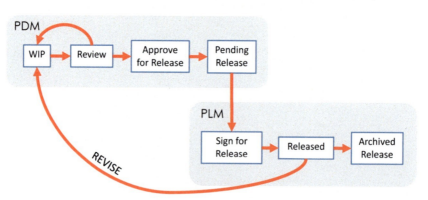

Business strategies may require both PDM and PLM software tool sets in order to cover the complete WIP (Work in Progress) to Archived Release process. Combined PDM and PLM tools can perform the entire WIP to Archived Release process.

A PDM/PLM database, capable of managing 3D data sets, works best in combination with its native CAD software suites, i.e. NX and TeamCenter, CREO and Windchill, SolidWorks and EnterprisePDM, Catia and ENOVIA.

At this point, for 3D data set management, stick with the PDM built for the CAD software. CAD software vendors are currently wrestling with integrating multi-CAD native formats into a collaborative environment.

As PDM 3D data sets shall be accessed via secure web portals, so also should PLM data be accessed.

Document vs. Data Management Approach

Most likely, your organization has a software tool in place that releases digital files (e.g. 2D PDF drawings) and plays the role of the PLM database. That's great! However, adding 3D models into a system meant for two-dimensional data release presents an additional layer of complexity.

The nature of 3D model definition requires files to be intimately connected. With a Model + Drawing product definition, it is necessary for 2D drawings and their parent 3D model to remain connected, as the 3D model and 2D drawing are associated. In addition, parts that have mating interfaces can have references to the originating interface feature. For instance, one part may control the hole pattern of a mating part. This creates a master-slave relationship between the two parts. Eliminating the associativity between parts severs the master-slave relationship. When a master-slave relationship is purposely created, time savings can be gained when the master is modified as the slave relationship allows the CAD software to automate the change within the slave part.

Historically, businesses have not had the hardware resources, data management techniques, nor software tools to maintain complex inter-part relationships when releasing their product data. As a result, most companies mandate that users break or prohibit inter-part relationships upon product release, which diminishes the ROI of model-based practices.

Therefore, to maintain ROI on the model-based investment, it is crucial to keep bi-directional connectivity and allow data to flow from CAD software into the data management database and back again. Almost all PDM software tools provide for check-in/out of any digital file, but at this time, they do not seamlessly work with CAD software for which they were not built.

Model-based integration requires CAD and PDM tools to be in bi-directional sync at all times. Using CAD and PDM tools designed for each other allows the syncing process to be automated; otherwise you must rely on human labor to implement the bi-directional sync.

Using a database architecture, as opposed to storing files on a server, facilitates multiple uses of single-sourced data. This permits business entities the ability to package data in a number of ways as suited to the end user, while still pointing to a single data source for the content. Storing data in a database jives with the concept of a Technical Data Package (TDP), as you may want to send one TDP with some content to a vendor and another TDP with more comprehensive content to your customer.

 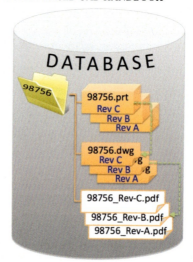

A server data storage architecture tolerates copying and pasting, creating multiple sources for the data. A database enables the ability to bundle different sets of data for different end items while bundling pointers to the source data sets.

3D AND 2D CAD MANAGEMENT
—*A Little Goes a Long Way*

Process vs. Tools

Bought any Ikea furniture lately? Seen the little Ikea blob guy? The blob guy and his pictures visually instruct you how to put together all those pieces of MDF lying in front of you on your carpet. The #2 Screwdriver and Elmer's glue are the tools needed to get the job done.

Without a process (step-by-step instructions), your finished product may look more like a jungle gym, rather than a bunk bed. Without correct tools, the bed may take much longer to build or may be improperly built.

Any data management system requires a process and tools that make the job possible. The tools used in a model-based environment are CAD, PDM, and PLM software. It is important for the desired data storage and release process to stand on its own, independent of software tool. Tools, soft or hard, do not define a process; rather they assist to implement it.

Managing CAD data requires the ability to manage both the drawing and model file while keeping a link between the two. For model-only data set type, it is only necessary to manage the 3D CAD files. In model + drawing data set type, it is necessary to manage both 2D and 3D CAD files.

3D CAD: The digital file(s) produced from any CAD software package that is the 3D model of an object. It may or may not include Product and Manufacturing Information (PMI) annotations.

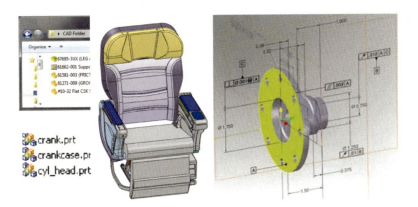

Examples of 3D CAD models — note these are only graphical representations (screen shots) of actual 3D data sets.

 2D CAD: The digital file produced from any CAD software package that creates a 2D engineering drawing, which is the source for either printed-paper or a digital 2D formatted file. These files may or may not derive from their 3D CAD parent.

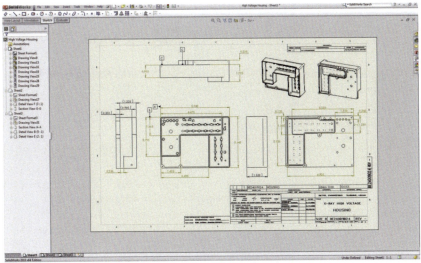

Example of a 2D CAD drawing.

Why Manage CAD Data?

As usual, the benefits are many, but not without challenge. Keeping tabs on both 2D and 3D CAD is critical in a model-based environment because the digital data IS the design and manufacturing authority. The only way to ensure accurate digital data delivery is to manage all the data set files in a digital database.

The two major objectives of CAD data management are:

1) **Common** CAD database repository—single source authority for the product
2) **Organized Control** of assembly and part files—systematic approach to product review, release and build

A well setup PDM will provide a single accurate authoritative source (accessible by all) to reference the product design documentation and offer it in a structured, well-organized manner. The PDM, in essence, is a multi-dimensional file cabinet.

When the two objectives of CAD data management are achieved, the following benefits can be realized.

- Efficient CAD sharing facilitates design collaboration over single or multiple physical locations.
- Single storage location of files facilitates team access.
- Rapid parallel design environment and concurrent engineering are enabled.
- Rapid customer review and product documentation delivery.
- Iterative changes are automatically archived.
- Team reviews the most recent design information.

The need to review CAD files starts much earlier in the design cycle than we think.

How do you control models?

In order to control models, determine your existing hardware release process. If one is not documented, then diagram it!

In order to modify the release process for model-based release, first identify the parts of the release process that are paper (2D drawing release only). Eliminate them or replace them with the appropriate 3D release. For instance, if redlining a physical piece of paper to review the product design, eliminate the printer and red pen and replace with a 3D reviewing tool.

It is not good to force processes that were generated for paper onto a digital creation and capture system. Keep the basic processes, but prune off the documentation methods that are no longer necessary. When you try to do both, you DO NOT gain efficiency. Therefore, it is important to constantly assess your CAD data management strategy for efficiency gains.

Document the 3D release process. Train users and give them quick reference aids in the revised released process. Without new instruction and coaching, humans will revert to what they know—which is how to release on a piece of paper.

Apply rules for users to use the PDM note or description area to make a quick reminder of what was done to a part or assembly when checking out, making a change, then checking back in.

Configuration management of YOUR digital data IS NOT a vendor solution.

Because your 3D models ARE your data, YOU must create the process to manage your data.

Remember, CAD software vendors are in the business of selling their software, not providing process definition. It is in CAD and PDM vendors' best interest to listen to your data management needs, but if you can't tell them WHAT you want, their software implementation will not be successful.

PRODUCT RELEASE CYCLE

Central to data management is the means to send each digital data file of a TDP through a structured process or workflow. Take a look at a basic workflow for releasing a product. Each file must step through a minimum set of states in order to be used as final product documentation.

For the model + drawing digital data type, both model and drawing must go through a revision process. As these two files are sent through the revision process, they are linked and should not be separated. For model-only, the model file is the only file to be released.

A majority of design projects rely on concurrent (parallel) engineering release, meaning two or more designers are working on interfacing parts at the same time. Concurrent engineering practices demand the ability to retrieve, reflect, and update the release state of the "goes-into" assembly or mating part. For example, if I am working on analyzing the bulkhead of a ship and discover a weak area between the keel and adjacent intercostals, I need to know the release status of all the mating structures so I know what I have the flexibility to change and what I do not. If one of the mated structure components is in *WIP* state, then I might be able to adjust that member; if not and it is already in *Released* state, then it is probably more efficient to change my bulkhead. It is extremely convenient when this data is accurately reflected in the CAD assembly tree hierarchy, PDM and PLM and all sourcing from a single variable.

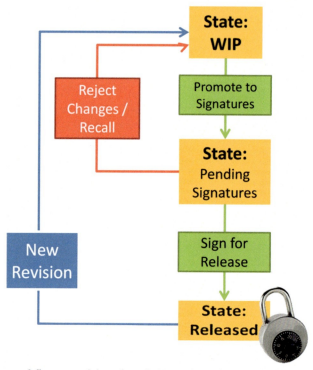

Basic workflow, requiring the minimum set of states (WIP, Pending Signatures and Released)

Examining the development of a helicopter illustrates the importance of live release state reporting when using concurrent engineering practices. The rotor assembly is undergoing the review to release process (perhaps taking several weeks). While this release is in progress, the project manager will not want to stop work on the development of the rotor harness and rotor structure interface designs to wait for the rotor to be finally released. It is necessary for the harness and structure designers to have access to the rotor 3D model geometry and its assembly interfaces, and to understand in what stage of release the rotor assembly is in (WIP, Pending Signatures, or Released).

A properly integrated PDM to CAD connection will allow for the most efficient reporting and integration of the rotor, and its mating structure and electrical harness data. Trying to manage in native CAD in a non-native PDM, will involve much manual manipulation and constant monitoring to ensure data status syncing.

How your organization and industry does business will govern what the product release process looks like. However, at a minimum, Workflows, States, and Signatures are needed to achieve robust hardware release architecture. Let's look at these three requirements in detail.

Workflows

A workflow describes a job-specific process and is a sequence of steps (flow chart) that define how PDM/PLM software tools should function. Traditionally, the industry calls these workflows ERPs (Engineering Review Processes). ERPs were implemented as computer database management entered the workplace to take advantage of a digital signature process.

Sometimes, ERPs expanded the signature process to include virtual tabletop reviews. This process is essentially the virtual computer equivalent of trapping the review team in a room, and having the team agree or disagree on the adequacy of the product and its documentation. A virtual tabletop strategy may or may not be a realistic solution for your company culture.

Many businesses still find the people-trapped-in-the-room process to be highly effective. Whether you decide on a personal or virtual review approach, at a minimum each 2D and 3D CAD file will need to have a workflow that starts with a WIP state, moves into a Pending Signature state and has a final release state.

Running alongside the hardware release process is most likely some kind of Engineering Change Request (ECR) system. Each business addresses ECRs in its own way and many have change management databases already in play.

In a model-based environment, because the 3D data must be changed, an ECR workflow must tap into the 3D data, meaning the ECR workflow may need to accommodate a manual process moving engineering change requests from the one database and into the PDM where the CAD data is stored. Depending on the size and culture of your business, a manual process may be feasible. However, for businesses with many users of the ECR and PDM systems, an automated link is recommended. This link is only required when the PDM and PLM systems are disparate.

States

A state is a software variable used in the PDM software tool in order to facilitate the digital data movement through the lifecycle and is not the same as lifecycle data status, or release status (Rev A, B, C).

Lifecycle Data: Information that describes the status of the part within a controlled procedure for development, review, release for manufacture, retention for maintenance, upgrades, retirement, and disposal. For example, the revision level and release date are lifecycle data that describe the administrative state of the part and change as the product moves through its review and approval cycle. Lifecycle data does not describe properties inherent in the part (such as the material type or manufacturer).[38]

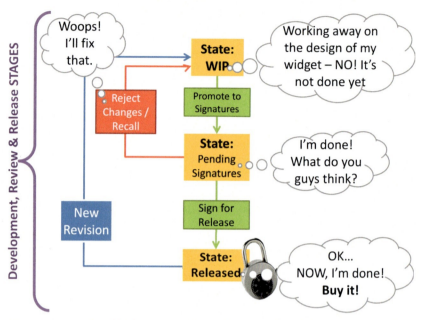

At a minimum, the PDM system will need 3 states: Work In Progress (WIP), Pending Signatures, and Released.

Signatures

Signature requirements are highly dependent on business practices for review and release. You may have required and/or optional signatures for every product or only a certain class of products. Most aerospace and defense projects require an engineer signature and quality signature, but may also require analysis, manufacturing, and management signatures to ensure all the appropriate stakeholders have reviewed the product to be released.

Your business practice of the desired hardware release will define how long the release process takes. The bottom line is that the more signatures are required, the longer it will take to get the product released. It is important to keep in mind that the length of time it takes to release a product generally has very little to do with the software employed for PDM or PLM and more to do with those who review and sign, their comfort level with the release process, and their ability to use the release tools.

Your organization can choose to do signatures by each individual logging into the system and applying his or her digital signature approval, or a single approver can approve for all. These are business decisions, and the software tool you choose should implement your desired signature process.

SIGNATURE AUTHORITY	DETAILED Part & Drawing	DETAILED Assembly & Drawing	MECHANICAL ICD	INSTALLATION DRAWING
DESIGNER	REQUIRED	REQUIRED	REQUIRED	REQUIRED
RESPONSIBLE ENGINEER	REQUIRED	REQUIRED	REQUIRED	REQUIRED
CHECKER	REQUIRED	REQUIRED	REQUIRED	REQUIRED
STRUCTURAL ANALYSIS	REQUIRED	OPTIONAL	OPTIONAL	OPTIONAL
THERMAL ANALYSIS	OPTIONAL	OPTIONAL	OPTIONAL	OPTIONAL
SYSTEMS ENGINEER	OPTIONAL	REQUIRED	REQUIRED	OPTIONAL
MANUFACTURING ENGINEER	REQUIRED	REQUIRED	OPTIONAL	OPTIONAL
QUALITY ASSURANCE	REQUIRED	OPTIONAL	OPTIONAL	OPTIONAL
MATERIALS & PROCESSES	REQUIRED	OPTIONAL	OPTIONAL	OPTIONAL

Example Signature Matrix

LIBRARIES AND CATALOGUES

Generally an organization has a quiver of parts that are commonly used by many designers in their product assemblies. Depending on company size, common part count could range from 100 to 100,000 individual parts. It is prudent that these parts be created once, checked once, released once, and stored in a controlled catalogue. A strategy of collecting commonly used parts provides engineers with the capability to call-up a desired part and reference it in their assembly with the confidence that the part is accurate as well as compliant with company modeling and document standards. Similar to the process for inventory management of that actual part, it is necessary to manage inventory of those parts' 3D representations.

Library: A collection of similar things (such as books or recordings).[39]

Catalogue Model: Any part or subassembly commonly re-used within an organization's products.

CAD designers need a repository of accurate parts, assemblies, materials, and notes to consistently reference when developing their parts and assemblies. Traditionally, we CAD designers have used libraries as a way to hoard anything we might ever use again. The problem is that if everything ever created is stored in the library, what items are active and relevant? It is vital that all elements housed in a library are active and relevant; any retired component must be shown as such within the library.

When you pick up the latest Pottery Barn or MicroCenter catalogue, rest assured the products in those catalogues are up-to-date and ready for sale. Regard CAD libraries as an active catalogue, where parts are plucked from accurate, updated database, and placed into an assembly with confidence.

Parts used by many designers, such as nuts and bolts, are a concrete example when thinking of library parts. However, it is important to also consider all data elements that contribute to the product design that might benefit from a common source. A good example of a non-concrete library is a material library. It is important not to have 5 different definitions for Aluminum 6061-T6.

Most CAD systems have unique material libraries that can be modified to your business needs and stored as a single data file or library for engineers to reference and apply to their 3D models.

Consider, PLASTICS R' US, a company that creates injected molded products. One designer is responsible for all product design, and then library implementation is not necessary. However, PLASTICS R' US hires a second designer who starts designing his own parts. The second designer enters a special density to represent the plastic that PLASTICS R' US uses to makes their parts. He modifies the material library locally on his own computer. The scenario has now been created where more than a single source of data exists for PLASTICS' plastic.

Multiple and most likely disparate data for the same material is sure to be a time-waster, which will ultimately hold up release processes, as the Checker must reconcile material disparities in addition to regular duties. An even worse scenario is that the disparity is never discovered and 1,000,000 parts are delivered using the wrong material.

In a model-based environment, a single source for material data is critical because of the hidden nature of model-based data. Most CAD systems do not "display" the material data in an obvious manner, and it would be simple to miss. However, PDM systems allow the display of title block information that reflects normally hidden metadata into a human consumable format, satisfying our human need to see the information.

It is easy to forget the libraries until the end of MBE implementation, but in reality, library setup must be accessible to users on day one.

In addition to material and catalogue libraries, most likely your business will benefit from setting up common note templates that all users can access to sustain consistency throughout engineering notes on similar products. There are several library types that benefit an organization.

Each organization will have different groupings of similar items that are defined in their CAD software suite as libraries. The following is a list of minimum libraries required to accomplish MBE.

Library Types
- Materials
- Standard Holes—Includes accurate thread callout detail to facilitate correct annotation
    - Clearance
    - Threaded
- Common Annotations
    - Notes
    - Weld Symbols
- Catalogue Parts and Assemblies

Imagine 10 people, who all have their own #10-32 x .500 inch long cap screw, #10 washer, and definition for steel. Will the assembly that collects those 10 peoples' assemblies have accurate BOM definition? Will it have many references to extraneous data? Will the file size be larger than necessary? The answer is that without specifically controlled libraries, you don't know the answers.

Most likely the parent assembly of those 10 subassemblies will have 10 separate #10-32 x .500 inch long cap screw digital files, 10 separate #10 washers, and who knows what steel those screws and washers are referencing. All these variations in the representation of the same screw and washer result in inflated assembly sizes, making it impossible to open a top assembly because it maxes out the available RAM on a client machine.

Implementing single source catalogue databases not only ensures accurate data sets for the common and standard parts, but will also control assembly bloat and extraneous database referencing (which reduces network traffic).

What are Catalogues?

A catalogue collects parts or subassemblies that are used more than once in a family of products. The catalogues might be comprised of a family of bearings, motors, bolts, or washers. In a 3D model-based environment, a catalogue part model can be assembled into your organization's native CAD assembly models and include accurate geometry (as specified from the supplier), attributes (material, color), metadata (part number, description, supplier), and annotations (dimensions and GD&T). A single catalogue part or assembly model is the single source for each catalogue item used in your organization.

Catalogue Types

1) ***Industry Standard Model Catalogue***—Commercially supplied part and assembly models generated from a specification managed and maintained by standards bodies such as AIA, ASME, SAE, and ANSI. A supplier or vendor controls the 3D model content. The format of the model shall be your organization's native format.

2) ***Commercially Available Model Catalogue***—parts and assemblies purchased from a supplier, such as: bearings, pumps, gearboxes, and holding clamps. A supplier or vendor controls the 3D model content. The format of the model shall be your organization's native format.

3) ***Common Model Catalogue*** —parts used by many in the organization, and are designed by the organization. For instance, a company that builds motors, may have common housings used in many different motor designs. The 3D model content and 3D product definition is controlled by your organization. The format of the model shall be your organization's native format.

Catalogue parts purchased from a supplier do not require annotations, as your organization does not build them from scratch, and therefore, product documentation is not required.

Additionally, when catalogue parts are instanced in an assembly, it is crucial that they are unambiguously constrained within the assembly. Otherwise, as with a mechanism assembly, as the mechanism moves, fasteners fall out of their connections, and the assembly product definition becomes ambiguous.

Before a product is complete, designers must integrate multiple components together into an assembly. Generally, standard or common parts are needed and unfortunately, these simple yet crucial parts are forgotten from assemblies or worse yet, purposely eliminated. However, when designers are given accurate libraries, their time is spent virtually testing those parts or materials to the benefit of the product design, rather than creating the parts.

Failure to implement a catalogue model solution into a model-based design process generally results in multiple models representing the same part and no control over catalogue models within the design database (PDM and/or PLM).

What Catalogue Library should you use?

Commercially available and industry standard model catalogues are available from companies who specialize in cataloguing CAD technology and deliver models on demand in a variety of native and neutral formats. In the spirit of re-use, it makes sense to start with data that has already been created. If modifications to this data are necessary, then they can be made, but at least your organization doesn't need to start from scratch.

Some vendors provide these libraries on-demand or for one-time purchase. Commercially available models are often maintained by offering suppliers. Industry standard models (ASME, MS, AIA, etc.) are fully represented in 3D with appropriate metadata and most major CAD native formats. Sourcing these re-usable items from a third party who guarantees their accuracy saves your organization time and effort, and creates accuracy in the assembly models you create.

Commercially available and industry standard models will be modeled with varied quality, schemas, and maintenance plans. Catalogue models, data, and update procedures should be compatible with the business entity's CAD modeling schema. Most catalogue parts created from a standard do not change often, so their maintenance is minimal. There might be a supplier that you use that regularly is updating the geometry or metadata of their parts. These updates should be monitored, similarly to how the physical inventory is monitored, with quality verification of the vital virtual data.

Library & Catalogue Benefits

Creating and maintaining libraries of parts, materials, notes, and company standard items allows re-use of single source accurate data. Spending time on library creation and maintenance is never wasted and is critical to the success of MBE. Even if model-based documentation is a secondary goal, a quick yet robust release process is sure to be your primary goal. Libraries assist in a more time efficient product release process.

Including libraries such as: catalogue models and materials, not only supports Model-Based Engineering and Digital Product Definition, but also have a number of additional tangible benefits:

1) ***Reference to a sole source:*** The instances in the assembly are just that—instances of the authoritative source, but have full interoperability with native CAD formats.
2) ***Ability to visually analyze*** and scrutinize bolted connections or inter-related parts for interference checking
3) ***Guaranteed modeling***
   a. Industry standard models are guaranteed to be modeled to the standard.
   b. Commercially available models are guaranteed to be accurate by the vendor.
   c. Common models are updated and flagged in assemblies when revisions occur.

SOFTWARE TOOLS

There is a plethora of software tools making it possible to create 3D models, translate 3D geometry, and manage digital data in a live collaborative environment. Each software tool has advantages, but your organization must choose a software suite suitable for their business practices. This is a tough one, and it may not always be obvious which tools are the best match for your business practices.

Begin with assessing how software tools perform when executing your business practices, using the following key parameters.

1) ***Ease of User Interface***—*Ability of users to execute the desired workflow.*
2) ***Ease of Administration***—*Capability of the administrative tool to create, manage, and improve workflows*
3) ***Robustness of Software***—*Ability of the software tool to accomplish and maintain your organization's desired data management scheme.*

It is important that your business entity select, implement, and use the right tool appropriate for any given stage in your model-based framework. A Model-Based Enterprise is best served when the software tools are matched to each stage of the job at hand, whether the stage is CAD creation or facilitating CAD interoperability. Conversely, it is important that you don't get hung up on the tools, for the tools will evolve and sometimes change.

First, identify what you want to do—then decide which software tools best fit your needs. Don't be surprised if you have a big software suite, but still need a third party software to provide precise geometry and PMI translation and validation. It is acceptable to use software outside your software suite if it fills a need within your model-based process framework and provides benefit to your organization.

Create a software tool roadmap that matches your business practices and product suite.

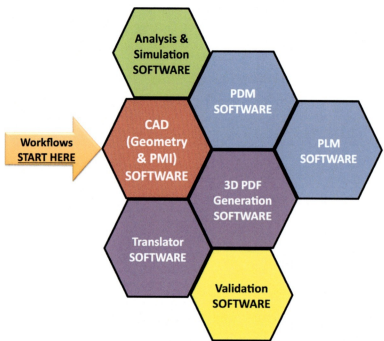

Software Tool Roadmap Template—Tools will generally be interconnected non-serially.

In addition to 3D Model Creation, your outfit may also need additional software tools, which assist in the following functions:

- 3D Model Analysis & Simulation
- 3D Model Translators
- 3D PMI Creators
- 3D PDF Generators
- 3D Model Validation
- PDM
- PLM
- Model Cataloguing Services

The interoperability of third-party software tools to your native CAD suite is another important task to keep on the MBE Implementation Team's radar. The inter-connectivity of software tools, whether automated or manual, determines the efficiency of your process. Failure to have a strategy that connects the software tools can put up major roadblocks in the release cycle.

Choosing the right tools for your business is up to you. The most important goal when selecting software is usability. If a tool is easy to use, it will get used the most. If a tool is hard to use, it won't get used. So no matter what the tool's capability, if it isn't used, it isn't doing you any good. As an example, higher end CAD software has a ton of capability when it comes to geometry creation, but if all the more complex features are too difficult to use, they won't get used—and you might as well not have paid for them. Often it is better to go with the path of least resistance, assuming the software still offers at least 90% productivity base. For the other 10%, you will need to be clever about how that work gets done.

IT REQUIREMENTS

CAD creation software requires a high level of collaboration with your IT department. Computer hardware selection must be a collaborative effort between the IT department and the MBE implementation team. It is a mistake for CAD, PDM and PLM experts to select software tools that may drive increase in network use, without corporate IT input. Inversely, allowing the IT department to select CAD hardware without consulting experts in CAD, PDM and PLM will often result in hardware that is ill suited to the job. Managing these two groups will require a balance of technology and budget.

For example, often deployed to CAD developers are computers with graphic cards that are incompatible with the CAD software and have insufficient RAM. Because operating budgets are a reality, buying top of the line hardware isn't always a realistic solution. Therefore, it is important to focus on the major

IT requirements and work side-by-side with the IT department to facilitate the optimal hardware solution.

Hardware

The four most important requirements for hardware used to create or view 3D CAD models are:

1) **Graphics Card Compatibility**—This doesn't necessarily mean the most expensive or the latest; it means it is compatible with the CAD software that runs on it.

2) **RAM (Random Access Memory)**—Assemblies and parts with complex geometry require fast access to data; RAM is the first line of action to get this. Max out RAM on laptops and get no less than 16GB for desktops. Given the ever increasing capabilities of CAD packages, and a typical life expectancy of a corporate computer of three to four years it is recommended that you put as much RAM in your machine as your team can reasonably justify.

3) **Fast data rate hard drives**—Solid-state hard drives are a great boon in this area. Not only are they mechanically robust, but also their data access rates (file copy, write, and open speed) are faster than a mechanical hard drive.[40] They are worth the extra cost.

4) **Single purpose CAD machine**—Many company security protocols require interrupting applications by accessing the hard drive and consuming RAM. If possible, the desktop CAD machine that does not leave the physical office can have minimal security images and only run applications critical for CAD development.

When CAD first came into being, the cost of a CAD capable machine was over three thousand dollars. Today's CAD capable computers are less than $3000 per unit and a viable commodity when compared with the cost of your engineer's labor.

The recommended working environment for CAD designers is one desktop computer with the best graphics card that matches the native CAD software, a solid-state hard drive (128 GB is generally plenty) and at least 16GB of RAM. Software loaded is limited to CAD needs only: CAD software, PDM, a screen capture tool, email and Power Point or the like to create charts.

In addition to the desktop, CAD designers need a laptop to take to meetings, on the road, or grabbing to show their boss their latest 3D ideas. The laptop needs only medium graphics capability and 4-8 GB of RAM. To keep it mobile, nothing larger than a 15-inch screen size is recommended.

A note about processors: most CAD software does not take advantage of multi-thread processing. So more processors in a single machine will not necessarily give you any more bang for your buck, but completely separate machines are a boon.

CAD software vendors recommend and certify particular computers. Stick with their recommendations, as the software will run most efficiently on the certified machines. This choice contributes to an increased ROI from your chosen CAD software suite.

Providing tools to assist, rather than hinder, a CAD designer's job is compulsory. It is not acceptable (given the expense) for an engineer to sit and play solitaire while waiting for a computer to crank through saving or rebuilding a part. This is mostly a problem for analyst machines cranking through Finite Element Meshes, but because model-based approaches increase the quantity of the data (i.e. more bits), computers spend more cycles cranking through complex geometry features and inter-part relationships. This data is all great stuff and enables a model-based environment, so rather than limit it, just buy good computers and buy several if one doesn't do the trick.

Non-CAD Users

For a model-based environment, non-CAD users will need to view model data. Make sure the viewing software and hardware are compatible with one another and create an environment that will not frustrate a reviewer. It is possible that the 3D model, view, review, and/or translate tool will require similar resources as the CAD machine. IT should be aware of these needs so users have access to correct hardware as needed for their job.

Again, computers are cheap compared to labor.

Visualization Software

Not all members of a team will need a license of the native CAD software to simply view the 3D model and associated data, but they will need a software tool and supporting hardware that facilitates their 3D model review function.

Visualization software is available as part of the CAD software suite (e.g. eDrawings, Creo View Express, JT2Go). Visualization software will run on computers with less capable graphics cards and generally access a 3D model format that is smaller in size, meaning large assembly models can be viewed in the visualization software because the full blown CAD creation capability is not accessible. Additionally, third-party software is available to view STEP and 3D PDF data that can be used in addition to or in place of the native software viewers.

Any member of the product team should have access to the viewing software in order to quickly view 3D model data sets. Employing visualization software enables managers, manufacturing, quality, configuration management, purchasing, and chief engineers the ability to view the source 3D Models. The benefit is that a long list of team members has access to view the source data.

When other team members have access to visually inspect the source data, it prevents CAD Designers from spending time on creating derivative graphics (power point charts) to explain the design, giving the designer a 3D accurate communication tool to help convey the design concept.

Realizing benefits from engaging visualization software will require culture change from non-CAD users to review the data in 3D. See Chapter 1.

Network

A fast network is critical for most business operations; however when you add in CAD data management and collaboration, demands on the network infrastructure are compounded. IT infrastructure will need to accommodate the increased network traffic when users check-in/-out assembly files, which are generally quite large with file sizes of 100-1000 MB, or greater. Analyze the file size of your company's assemblies to determine average network traffic and size the network accordingly.

Following best practice, an individual designer will most likely check-in/-out 100-1000 MB sized files 1-4 times per day.

Maintenance Plan

Software upgrades are inevitable. Sticking your head in the sand will not make them go away. Most companies plan software upgrades for CAD related tools every two years. This interval is too long. At a minimum, major upgrades occur once per year and at most service packs are released quarterly.

Whether your company policy requires you to lag behind an upgrade or you are free to upgrade as soon as released, set up regular upgrades, as the benefits of staying current are many. If organizations that inter-operate all stay current, then version incompatibilities will be fewer.

Also of note: most vendor software purchase contracts are really more like a rental rather than purchasing outright. Because software is constantly in improvement mode to keep up with customer demands or to stay current with the hardware it operates on, it is best to plan for upgrade intervals no sooner than quarterly and no longer than annually.

[5] General Model-Based Schema

Models created without a common organization of the 3D data set will drive up file sizes and result in assemblies and drawings that will not open. In contrast, organized models built using good modeling practices will regenerate without errors. Models that have regeneration or rebuild errors wreak havoc in their Next Higher Assemblies (NHA). A model created with consistent metadata and starting templates will flow easily throughout its model-based lifecycle.

Every company must have a model schema that defines how a CAD model should be structured.

Model Schema: An organization, grouping, naming convention, and direction for completeness of models, annotations, attributes, and metadata included in the model. These basic building blocks facilitate accurate data exchange and archival of TDP Type: 3D.

Next Higher Assembly (NHA): Identifies the assembly that the part or subassembly inhabits.

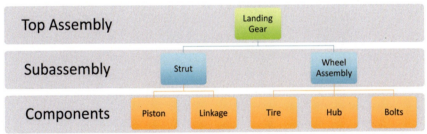
Drawing tree identifying the top assembly, subassembly, and components.

Solid Modeling Philosophy

The over-arching goal of a model-based 3D solid modeling philosophy is to produce a master definition of products that are converted into derivative data sets only when necessary. Greatest gains will be seen from a model-based environment by moving away from 2D and 3D drawings and leveraging the benefits of a model-only approach.

3D Drawing: A 3D solid model with complete annotations that unambiguously reflect all solid model dimensions as are typically found on a 2D drawing.

NOTES (UNLESS OTHERWISE SPECIFIED):
1. DESIGN MODEL 67801.STEP IS REQUIRED TO COMPLETE PRODUCT DEFINITION.
2. OBTAIN DIMENSIONS FOR ALL UNDIMENSIONED FEATURES FROM THE MODEL.
3. ALL DIMENSIONS OBTAINED FROM THE MODEL ARE BASIC UNLESS OTHERWISE SPECIFIED.
4. TOLERANCE FOR ALL UNTOLERANCED SURFACES = ⌓ | .020 | A | B | C |
5. INTERPRET DRAWING IN ACCORDANCE WITH ASME Y14.5-2009, ASME Y14.41-2012 AND ASME Y14.100-2004.

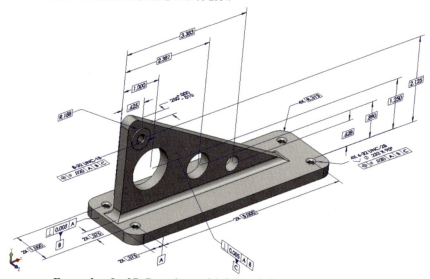

Example of a 3D Drawing, which is a fully annotated model.

At this point in the book, I feel as if I am beating a dead horse, but in case you have skipped ahead to this chapter, here is a synopsis of the solid modeling philosophy from MIL-STD-31000A.

- **All required engineering information** is communicated to everyone and originates from one source.

- **The goal of a solid model is to create and provide** all design and detailed information to downstream users.

- **For effective communication**, consistency is crucial in the creation and presentation of each product model.

- **Maintaining integrity of model data is the responsibility of all engineers and designers and anyone else** who may add or change the model during its creation and revision.

In order to facilitate the solid modeling philosophy required by MIL-STD-31000A, a set of rules of how models will be created and documented is required. This is called the model schema.

Solid Model: A mathematically accurate three-dimensional representation of an object. A solid model has thickness and when calculated, results in numeric values for mass, center of mass, and Moments of Inertia (MOI). A solid model is not a collection of disconnected, unstitched surfaces.

 Surface Model: Mathematical or digital representation of an object as a set of planar or curved surfaces, or both that may or may not represent a closed volume.[41]

THE SCHEMA

Chapter 2 provides insight into which models constitute the authoritative source model files. In the development cycle, there are four models that are in play: Design, Product, Manufacturing, and Archive. The Design model instigates the process and flows data into the Product, Manufacturing, and Archive models subsequently.

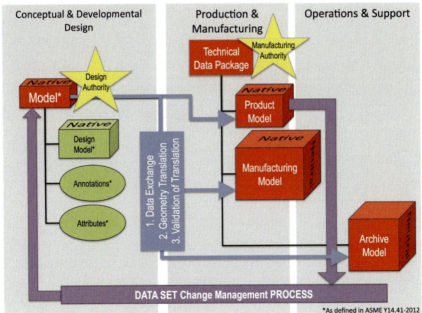

Model Road Map

As defined in Chapter 2, there are two authoritative sources within the product lifecycle: Design and Manufacturing. Ideally, one model schema covers all data and organization needs to accommodate both the Design and Manufacturing authorities. However, if necessary, an additional manufacturing model schema can derive from a design model schema. The two schemas should not be mutually exclusive and contain common organization where appropriate.

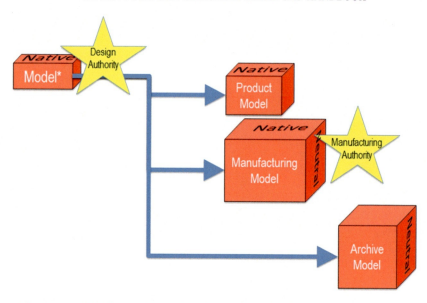

A look at model files and how they derive from the design authority model.

General Model Based Schema Requirements

General schema requirements address those schema elements common to both part and assembly models. Drawings are also covered under the general requirements, but are unique because of the nature of model-based documentation in the model + drawing data type. Metadata and data management techniques presented in the general schema also apply to drawings.

{recipe} —— Easy & Yummy MBE Cake

1. Combine Nuts and Bolts together in a bowl – Set aside
2. Melt Lubricant into Brackets
3. Attach brackets together using Nuts and Bolts
4. Add other components as necessary
5. Place assembly into pan.
6. Bake at 500° F

{ingredients}
1 c. Nuts
½ c. Bolts
2 large Brackets
1 tsp. Lubricant
(optional) Other Components

A model schema is the recipe required to produce consistent model-based solid models.

How does the schema get implemented?

The term "standard" is used in many different contexts. Experience with many process implementations has shown that one person's "standard" is another's "guideline" and still another's "nemesis." In this chapter we will present a set of rules and best practices that define 3D models. It is your job to tailor them to best fit within your company's business practices.

Simple or small projects might implement the schema quickly, while large complex projects will require in depth planning, roll-out dates and MBE Implementation team members, as described in Chapter 3. Small projects may get away with manual operations, such as the 3D checker manually verifying compliance to the model schema. Large projects will most likely require automation tools, such as bulk data translation and validation of the model schema.

Why do I need a schema?

Accurate, well-organized data is easier to consume and archive. If you don't understand the data presented to you, you can't work with it, much less understand it. A model schema provides basic rules and best practices to build accurate, well-organized 3D data sets. It is also important for the standard CAD rules to stand independent from CAD software implementations.

Most CAD modeling software is capable of supporting a model-based environment. What has become increasingly frustrating to users is not whether a function is possible, but rather how the user would like to accomplish that function. For instance, to provide enough information to produce a motor mount bracket, is it enough for the solid model geometry to define the motor mount shape with the model-only or must the dimensions be replicated on a 3D drawing?

In the same way that the ASME Y14.5 standard is used as the 10 commandments of drawing creation, so should ASME Y14.41, MIL-STD-31000A and this handbook be the 10 commandments of Model Based Engineering (MBE). It is important that industries use common techniques to define their products, to give suppliers common ground when producing products for several customers. When each customer has unique documentation requirements, economies of scale cannot be leveraged, and the cost will be passed on to your customer.

This handbook presents a comprehensive set of CAD rules in an effort to organize the design data and communicate the design from the designer, onward to procurement, and onto the shop floor. A CAD model schema will add order into the collaboration and interoperability chaos.

Technology Elements that Enable Schema Success

There are a few key software implementations to focus on because these are the biggest bang for your buck for managing consistency throughout the models in your organization. Also of importance is that the following CAD and PDM/PLM tasks be implemented ASAP.

1) ***Templates for parts, assemblies, drawings, and drawing formats**—Commonly called "start parts"*

2) ***Customer defaults within each CAD, PDM, or PLM**—Commonly stored in configuration files and unique to each software CAD package.*

3) ***PDM/PLM Roles and Groups**—Strategy for implementing data management techniques*

Terms and Definitions for the Schema

The presented terms are deliberately not in alphabetical order to best relay their meaning. Read and understand each term and its definition before moving to the next one.

Part: A single hardware item.

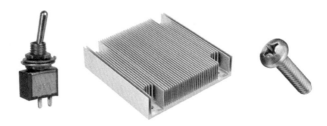

Parts of an electrical enclosure.

Assembly: A collection of parts that can be non-destructively disassembled.

Electrical enclosure assembly where the part to part connections are made via removable fasteners.

Model: A combination of design model, annotation, and attributes that describe a product.[42]

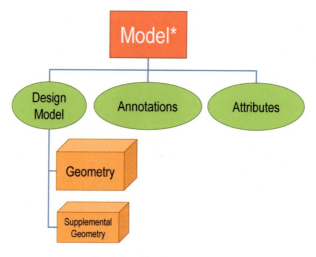

*As defined in ASME Y14.41-2012

The term Model has many definitions. For this handbook, we will use Model as defined in ASME Y14.41.

Design Model: The portion of the data set that contains model geometry and supplemental geometry.[43] Supplemental geometry is optional and not required in each Design Model.

Model Geometry: Geometric elements in product definition data that represent an item.[44]

Example of a model with only the geometry displayed.

Supplemental Geometry: Geometric elements included in product definition data to communicate design requirements but not intended to represent a portion of an item.[45]

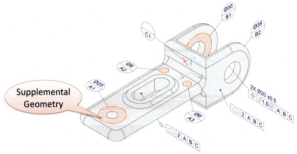

Red crosshatched zones drawn on part are examples of supplemental geometry. Graphic courtesy of Advanced Dimensional Management LLC[46].

Annotation: A graphic or semantic text entity that describes the dimension, tolerance, or notes of a particular feature in the model. It is stored and visible in the model as an exact and permanent form of digital product definition.

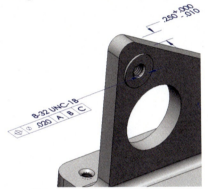

Example of an Annotation

Attribute: A dimension or metadata of a particular feature that is not visible in the model and is accessed via model query.

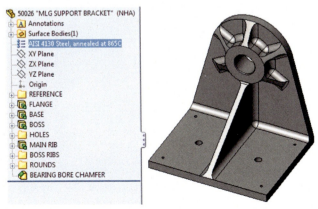

A material is an example of an attribute.

Metadata: Data that supports the definition, administrative, or supplemental data package. Metadata includes all relations, parameters and system information used in a model. This data resides at the model and feature level.[47]

Axonometric View: A model view that is primarily used to display annotations, where the surfaces of the model are not perpendicular to the viewing direction.

Axonometric oriented view.

Additive Manufacturing: Process of joining materials to make objects from 3D model data, usually layer upon layer, as opposed to subtractive manufacturing methodologies.[48]

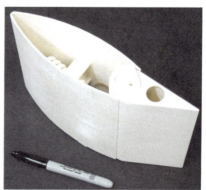

All-in-one flexible rudder additively manufactured (3D printed) using FDM process with ABS M30 material. Rudder designed and created by Walter Holemans, President of Planetary Systems Corporation[49]

INTERPRETING THE RULES

All rules and best practices listed in this chapter apply to both parts and assemblies. Some rules derive from requirements in MS-31000A or ASME Y14.41; others elaborate on these standards to provide a deeper definition to MBE.

As the model schema is described, several icons will be used to enhance the reader's understanding of the rule and assist in how and when to police the rules.

The following legends describe symbols that provide the reader with additional insight into the rules and best practices described in this handbook.

Data Set Type Symbols: Used to depict if a rule applies to the Model + Drawing type or to Model-only.

MODEL & DRAWING
• Authority for 3D Geometry = Model
• Authority for GD&T = Drawing

MODEL ONLY
• Authority for 3D Geometry = Model
• Authority for GD&T = Model

Process Clock: The clock represents the life of a product from development to release. Some rules presented are necessary at the creation of the CAD file, and some can be implemented at the time of release. The Process Clock is used to identify when a rule must be implemented in the release cycle.

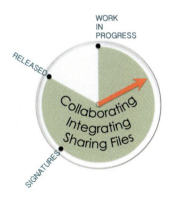

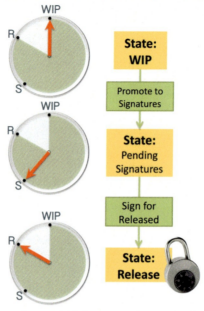

Method Stamps: Many times multiple approaches are valid and are designated as 1 through 3, where 1 is preferred over 2, and 2 preferred over 3. When a red X is used, this approach should be stopped immediately, regardless of prolific use.

CAD Rules

When the following CAD rules are included in the CAD (part, assembly, and drawing) "start part" templates, then the execution of these rules becomes easy. If these rules must be applied to a model after its inception, then compliance with the rules will be more challenging as these rules are fundamental to CAD model creation.

RULE: Model Representation

A unique solid model shall represent each unique component in an assembly.

RULE: Scale

All models shall be modeled at FULL scale: 1/1

RULE: Units

An organization will select Metric or English units as the default unit of measure. Exceptions to the default units require additional part, assembly, drawing, and format templates created with the alternative units.

For MIL-STD-31000A TDP delivery, English units shall be used, unless otherwise stated by the contract or purchase order. If metric units are used, they must conform to FED-STD-376[50].

RULE: Metadata

Metadata elements shall be uniform throughout a company for: parts, assemblies, drawings, supplemental TDP documents, PDM and PLM variables. Exact element names and descriptions will vary for each organization. The exception is when metadata element names must be adjusted to satisfy customer requirements. These exceptions shall be by project only.

MIL-STD-S31000 (B.7.8) Metadata

Critical Data Elements List

ID	Element	Description
P1	ALT_MATERIAL	Use as needed to define alternative materials
P2	BOM_NOMENCLATURE	Nomenclature for use in the parts list
P3	CAGE_CODE	Company Cage Code
P4	CHK_DATE	Checked date
P5	CHK_NAME	Checked by
P6	CONTRACT_NO	Parameter for contract number
P7	CONTROL_ACTIVITY	The organization code of the group with the control authority
P8	CRIGHT_DATE	Copyright year
P9	DATA_RIGHTS	The code, which identifies the rights status of the information on the Model or contained in the file identified by this record.
P10	DESIGN_ACTIVITY	The name of the design activity whose CAGE is assigned to the Model.
P11	DIST_CODE	The distribution statement code letter (A, B, C, D, E, F, or X) of the model identified in this record.
P12	DOC_TYPE	A code entered that identifies the class or type of engineering model (e.g., product drawing, parts list, wire list, safety data sheet, etc.).
P13	DRAWING_NUMBER	Drawing Number
P14	DWN_DATE	date the drawing was created
P15	DWN_NAME	Name of the drawing author
P16	ENG_DATE	Date of the approval engineer's signature
P17	ENG_NAME	Name of the approval engineer
P18	EQV_PART_NUM	USE if needed to define an alternant part for this item

For Technical Data Package (TDP) delivery in compliance with MIL-STD-31000A, critical data elements in MIL-STD-31000A, Appendix B, Section B.7.8 are required.

It is of utmost importance that all metadata is consistent throughout the enterprise. Each document MUST start with correct data or it MUST be ADDED downstream. Inconsistent metadata is very easy to create. Inconsistent metadata is the aspect of MBE that slows down and sometimes eliminates all enterprise software tool benefits the quickest.

> *How long will it take you to make a loaf of bread without a recipe? Longer than if the recipe were on your counter.*

Most of the metadata elements will be controlled via PDM and/or PLM. However, most CAD designers will be familiar with metadata as represented in the CAD design software as custom properties. The following tables represent the minimum metadata elements required. If compliance with MIL-STD-31000A is required, refer to Appendix B.7.8 for full metadata definition.

Element / Variable Name	Description
Number	Unique identifying number applied to each document. Follow numbering guidelines in the next section. Used in the Parts List.
Description	Text applied to each document that describes the data file to which it is attached. Some companies have standards for how the description is to be written. Used in the Parts List.
Revision	Appropriate revision identifier per company release process. For examples: A, B, C Letter Series, WIP: in development prior to release stage or C-WIP: Revision C in process of revision to Revision D.
Revision Date	Date release occurred.
Units	Identify Metric or U.S. Customary
Material	Material identification reflected from the material library. Only used in the part definition.
NHA	Identify all Next Higher Assemblies (NHA) that this data set feeds into. Configuration management often call this effectivity.

Minimum metadata elements set for each digital file (Table 1 of 2)

Element / Variable Name	Description
Company	Name or identification for responsible company.
CAGE Code	Company cage code per Federal Cataloging Handbook H4/H8.
Project Number	Contract Number or identification
Originator Name*	Name of the individual that created the document.
Approver Name*	Name of the individual that approves the file for release. This is the minimum signature that is required. Some organizations may require many approvals for release of a data set.
Approver Function	Description of the function of the approver. Some organizations may require many functions to approve a single file.
Approval Date	Date of the approval
TDP Type	Identifies whether this data set is Model + Drawing or Model-Only
Data Rights	Identify the appropriate requirements for restriction access, availability, proprietary data, or use of this digital file
Classification	Identifies whether this digital file is UNCLASSIFIED or CLASSIFIED with nomenclature per DOD Manual 5220.22-M

*Names are typically identified as First Initial and Last Name (All Caps)

Minimum metadata elements set for each digital file (Table 2 of 2)

RULE: File Numbering

In the aerospace and defense world, agreeing on a part and assembly numbering strategy is similar to religious disagreements. Entrenched standards have been in use for generations to create part numbers that tell the story of the part. This practice is commonly referred to as smart numbering.

However, because the data for a part now can be accessed and viewed in a number of ways, the traditional part number becomes less important. Searches allow for name and description searching, which with very large, interrelated data is highly valuable, as it saves times over manually searching for your desired digital file.

Because a digital database allows searchable access to retrieve information based on a number of variables—e.g. material, shape, size, description, where used—the significance of the part number is no longer as essential to document communication as it once was.

Having a unique identification number is now the more important characteristic of the data file.

Presented are two methods for numbering. Although one or the other may still be controversial, just pick one and stick with it.

 Auto Numbering: The PDM or PLM shall serially generate auto-numbers so that every file has a unique identification number.

 Semi-Smart Numbering: Every File has a unique identification number but also has a prefix that identifies the project.

Semi-Smart Number = "Project Number"–"Serial Number"

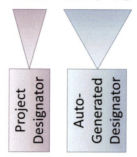

Example of a semi-smart numbering scheme.

RULE: File Naming

The filename is the actual name given to the data file in question. For instance, a part file may be named "release_pin.prt" with a number of "33331". What should the filename really be? There are two options.

① Filename equals Number: Most CAD systems can be set up to display number and/or description in the native hierarchy tree. Note that not all translators carry the metadata to the neutral format hierarchy tree; in these cases they leave only the filename in the hierarchy tree and lose the description. If looking at an assembly of parts, and the parts are only identified with numbers, the engineer on the receiving end of the model translation is confused because the descriptor is missing (meaning no description in the filename).

> Filename = Number + Extension
>
> | 33331.prt | 35432.dwg | 40001.pdf |
>
> Description is Metadata
>
> RELEASE PIN

Example of a filename equivalent to the Number variable.

Name and Number Associativity: Models and Drawings should be associated through attributes and metadata and not identified as a group through their number. It is a common practice for the number + extension (.prt, .dwg and .pdf) to have the same number with the extension used as the method to make the filename unique. This method of filenames with the same number, yet different extensions dilutes the importance of the unique number and is not recommended.

② Filename equals Description + Number: If your business practices require this method of file naming, it is an additional piece of data to ensure it is synced, as it is not common practice for CAD systems to auto-generate a filename from the description metadata element. If the software tools available do not maintain number and description metadata values through the required data translations, then a method of description-plus-number assists human consumption of the model, post translation.

Filename = Description + Number

release_pin_33331.prt

release_pin_35432.dwg

release_pin_40001.pdf

Example of a model filename equivalent to the Description and Number variables concatenated together.

Drawing Tree: A drawing tree is the product structure of an integrated assembly. It is an important systems engineering tool that helps keep a project in sync with the parts of the puzzle that create the final product. To maintain order, part, assembly, and drawing nomenclature (Number and Description) should match a product drawing tree.

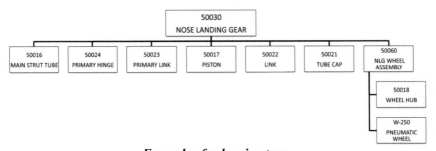

Example of a drawing tree.

RULE: Data Package Identification

Each data package shall have a collector database item "paper-clip" that binds those files together and identifies them as a Technical Data Package (TDP). PTC Windchill calls these "paper clips" WT parts. Dassault's Enterprise PDM requires an actual digital file to be the collector rather than an item in the relational database. That file can be as simple as a text file listing Number and Description, or as complex as an auto-generated Engineering Change Order (ECO) form auto-populated with solid model and PLM metadata.

TDP Package File Hierarchy Example

RULE: Dimension Value

Model all geometry to nominal dimensions. Define assembly constraints as nominal.

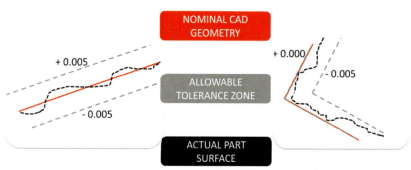

Examples of geometry modeled to nominal dimension and its surrounding tolerance zone.

RULE: Model Precision and Resolved Dimensions

ASME Y14.41 defines model precision and resolved dimension and their implementation practices. Refer to ASME Y14.41, sections 9.0 and 10.0 for complete rules for applying dimensions to models in compliance with Geometric, Dimensioning & Tolerancing (GD&T) digital data practices.

Model Precision: The highest number of significant digits set in the CAD model document.

Resolved Dimension: The dimensions shown on an annotated model or drawing. Resolved dimensions are representations of model values rounded to a desired precision and will have equal or less significant digits than the Model Precision. [51]

X.XXXXXX	2.000817	Model Precision
X.XXXXX	2.00082	
X.XXXX	2.0008	
X.XXX	2.001	Resolved Dimension
X.XX	2.00	
X.X	2.0	

Significant digit table identifying example of a resolved dimensions in comparison to the model precision.

RULE: Dimension Visibility

In both Model-Only and Model+Drawing data set types, the model geometry serves as the master of the dimension value. Because dimension values will be obtained by querying the model or automatically interpreted by another software, it is not necessary to display a dimension that complies with default tolerance information given in the data set. Default tolerance information can be given via tolerance block, note, or metadata where appropriate.

RULE: Dimension Associativity

Per ASME Y14.41, a direct and permanent association to the originating model value shall be established.[52] Any resolved dimension or queried model value is considered a derivative.

 Never "fake" or edit a dimension.

"Faking" dimensions is most commonly practiced in drawing creation. However, never fake a dimension. Easy to say, hard to do. I know when you are under the gun, it seems that editing a derived dimension (not the source creating dimension, usually found in the feature sketch) will save you time, but it will only drive inconsistency into your data sets, creating havoc downstream. In addition to faking a dimension being extremely poor practice, it is a requirement of ASME Y14.41 to not fake dimensions.[53]

RULE: Tolerances

Because dimension values are interpreted directly from model geometry rather than "read" by humans, standard tolerance schemes (i.e. standard tolerance block plus/minus tolerance) are not appropriate for data sets defined through Model Based Definition (MBD).

The more appropriate and unambiguous method is to use a profile tolerance general note to define the default overall tolerances.

TOLERANCE FOR ALL UNTOLERANCED SURFACES = ⌒ .010 A-B C

Recommended note for default tolerance for parts defined per ASME Y14.41. Note courtesy of Advanced Dimensional Management LLC[54].

RULE: Notes

Notes are generally unique to each company, but save time and maintain consistency by having a common note library accessible by users. This is true for 2D drawings or model-only drawings. In some MBD implementation cases, such as SolidWorks to 3D PDF format, it may be practical to include default general notes as custom properties so that they are transferred to the text area of a 3D PDF, rather than leaving them as 3D graphic entities.

Notes required in model-only must always be displayed on the model when accessed via viewing software tool. Additionally, notes must remain flat to the screen while the model rotates beneath it.

The following notes are required for each model-based digital data set.

Note	Model	Model Only
DESIGN MODEL <NUMBER>.<extension> IS REQUIRED TO COMPLETE PRODUCT DEFINITION.	X	
OBTAIN DIMENSIONS FOR ALL UNDIMENSIONED FEATURES FROM THE MODEL.	X	
ALL DIMENSIONS OBTAINED FROM THE MODEL ARE BASIC UNLESS OTHERWISE SPECIFIED.	X	X
TOLERANCE FOR ALL UNTOLERANCED SURFACES = [Profile .010 A B C] – Create per ASME Y14.5 and desired tolerance	X	X
INTERPRET IN ACCORDANCE WITH ASME Y14.5-2009, ASME Y14.41-2012 AND ASME Y14.100-2004.	X	X

Notes required to define the product using model-based definition.

RULE: Holes

Holes consist of clearance and threaded holes. The attributes associated with a clearance hole shall be: diameter, tolerance, depth, or through. The attributes associated with a threaded hole shall be minor diameter, major diameter, thread pitch depth, or through hole.

Hole data is best maintained as a library so that the data is single source and referenced among CAD creators. CAD modeling software often calls this library a "Hole Wizard". As with any library, as detailed in Chapter 4, the Hole Wizard library should be controlled with a change process identified to modify any of the hole values or attributes should the need arise.

RULE: Fully Annotated Models

Dimensions displayed on a fully annotated model must agree and source from the model geometry.

Fully Annotated Model: A model that includes every dimension and tolerance value necessary to produce the part by visual inspection of the annotated model alone. Also known as a 3D Drawing.

Caution when using fully annotated models must be exercised either through manual or automatic validation of data, as conflicts may exist between model geometry and displayed annotation.

This handbook recommends not using fully annotated models (3D Drawings), as that strategy provides more opportunities for data to be duplicated and in conflict. However, customers may require fully annotated models, making the practice a necessary evil.

Preferred product definition strategy for model-only and model + drawing type definition. Note that dimensions for this part are accessible via exact data transfer or model query.

RULE: Dimensions, Tolerances and Notes on Commercially Supplied Standard Parts

Catalogue, industry standard or commercially supplied parts used in model-based assemblies do not require annotations (dimensions, tolerances, or notes) as defined by MBD rules. The reason for this is that your organization is not responsible for defining or producing industry standard and commercially supplied parts. Recall that a commercially supplied part is purchased from a supplier and your organization is not responsible for defining it. Accurate geometry and metadata are the only MBD requirements for standard catalogue parts.

RULE: Default References

Default references refer to reference features in a model that help define the model origin, such as: coordinate systems, planes, and axis. Default references are different than ideal planes used as datums for manufacturing (Datum Reference Frames) and used only in product definition, not in solid model feature creation.

Although default planes can be used as Datum Reference Frames (DRF), it is not required that they be equivalent to the default planes in the solid model. In most real-life examples, the default model planes will rarely be used as the DRFs because the model creation process was started long before the part was analyzed for manufacturability. Reference planes that represent DRFs (e.g. A,B,C) may be added later as required by digital practice standards from ASME Y14.41.

Default Coordinate System, Planes and Axis: Every part model must have a default coordinate system, planes, and axis. Some CAD "new" parts automatically begin with default coordinate systems, planes, and axis at the model origin, and some do not.

The following default coordinate system, planes, and axis are required in the "start part" template of each part and assembly model. The coordinate systems, planes, and axis shall be designated as reference geometry and shall be available to the designer to reference during feature creation and component mate functions.

In cases where the CAD system does not provide default coordinate system, planes, or axis, start part templates should define default coordinate system, planes and axis, so that each user need not create them individually.

The following graphics illustrate examples of default reference coordinate system, planes, and axis.

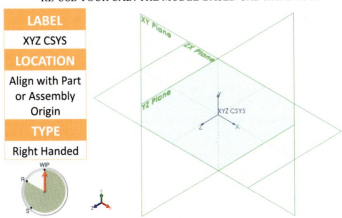

Default Coordinate System example.

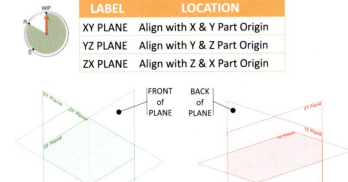

Default Planes example.

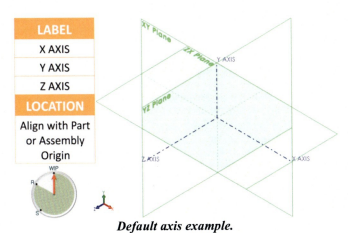

Default axis example.

RULE: Annotation Planes

There shall be three default planes (Front, Right, and Top) that may be, but are not required to be, coincident with the default XY, YZ, ZX planes.

An additional annotation plane is required to hold all notes on a single plane. The note plane shall be included in the part and assembly template, and shall not rotate with the part. The note text stays flat to the screen while the model is free to rotate, zoom and pan beneath it.

For annotations that do not conveniently lie on the Front, Right, and Top annotation planes, create Left, Back, and Bottom planes as needed.

Because there are not a great deal of CAD users implementing 3D annotations, CAD software vendors have yet to make 3D annotations particularly user-friendly. That being said, your organization will need to experiment with the best implementation methods for annotation placement and default annotation planes that work best in your CAD system and for your engineering culture.

Annotation Default Planes	Model	Model Only
FRONT (Parallel to XY Plane)	X	X
TOP (Parallel to ZX Plane)	X	X
RIGHT (Parallel to YZ Plane)	X	X
NOTES (Parallel to XY Plane)		X

Default Annotation planes for each TDP type.

CAD Best Practice

The rules presented up to this point are rules to be followed on each part and assembly CAD file, which stem from a standard or are required by any CAD system to be successful (i.e. default planes).

The exact implementation of best practices relies more heavily on individual CAD software capability, and therefore is presented as guidance for successful CAD modeling in a model-based environment. If suited for your organization, a best practice can certainly become a rule within your particular model-based implementation.

BEST PRACTICE: Logical Groupings

The purpose of creating groupings in a model is to facilitate model reviewing ease. Group together the following like components: reference components, features and patterns. Combine components and/or features that are to be suppressed/un-suppressed and/or hidden/shown together, such as fasteners. Logical groupings are required by MIL-STD-31000A.

Available CAD software implementations of groupings vary. Layers are the most traditional implementation of combining like elements within 2D CAD files. Some CAD software has the capability to utilize layers in a 3D model, and some do not. As layers tend not to be the user-friendliest grouping method, it is recommended to use a more intuitive way to group or collect features and geometry, such as using folders.

The following table describes groupings that must be included in the part and assembly templates (identified in yellow), as well as optional groupings that may be required depending on the product design.

	Folder Name	Part	Assembly
ANNOTATIONS	ANNOTATIONS	X	X
REFERENCE	REFERENCE	X	X
FASTENERS	FASTENERS		X
OPTIONAL	CABLING		X
OPTIONAL	LOFT SKETCHES	X	
OPTIONAL	ASSEMBLY CUTS		X

Yellow highlighted rows are groupings required of part and assembly models.

The following examples illustrate the use of folders to collect a variety of elements within a part and assembly model. Required folders shall be common to all part and assembly files to gain advantages for automated display schema options. For instance, if all fastener and reference geometry are contained in a group labeled "FASTENERS" or "REFERENCE", then it is possible to set up automated macros which suppress or hide all fasteners and reference parts or geometry in an assembly.

Additionally, logical groups aid CAD users who might inherit or need to modify the part or assembly, as they provide a roadmap that shows how the geometry was created or how the components in the assembly were put together.

Example of logical groupings using folders to group like features in the part feature tree and like components in the assembly feature tree.

BEST PRACTICE: Display Schema Methods

Building on the concept of logical groupings to organize the model, also available in CAD software are display schema methods that facilitate an organized presentation of visualizing the model and its documentation (notes, tolerances, title block, etc.) These methods are useful to help organize an assembly into various configurations, such as stowed or deployed for a moving assembly. They can also be used when working on a model to achieve better clarity in a particular area of the product design. They can be used to create saved views, enabling required combination views from MIL-STD-31000A.

There are 6 different methods of displaying features of a part, or components in an assembly. They are:

1) *Hide/Show*
2) *Suppress/Un-Suppress*
3) *Filter*
4) *Change Dimensions*
5) *Transparency*
6) *Large Assembly Visualization*

1) Hide/Show Method: Turns feature geometry and component models on/off, hiding them from view. Hide does not remove feature geometry or component models from memory (RAM). Therefore this is not a method to reduce part or assembly load times. An advantage of hide/show over suppress/un-suppress is that it does not eliminate inter-feature or inter-part relationships. As they are not grouping-specific, hide/show methods are not necessarily meant to be permanent. Some CAD systems accommodate saving specific hide/show arrangements.

TIP: Hiding components in and assembly is not a method to manage assembly load times.

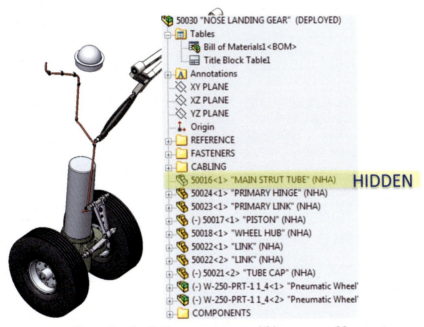

Example of a hidden component within an assembly.

2) Suppress/Un-Suppress Method: Turns feature geometry and component models on/off, removing them from view and the model. Suppress removes feature geometry and component models from memory (RAM). Bear in mind that suppressing a feature or a component will temporarily invalidate inter-part and feature relationships, because suppression turns off the feature or component, removing those items from the model as if they were no longer there.

TIP: Suppressing components in an assembly is the primary method to maintain explicit control over assembly load times, such as suppressing fasteners in an assembly or complex features.

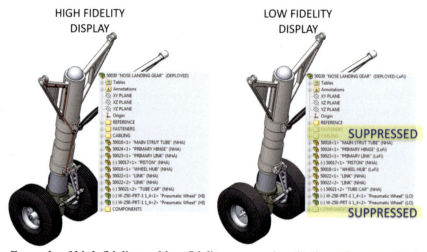

Example of high fidelity and low fidelity suppression display schema method.

3) Filter Method: Common grouping of features and annotations with buttons or drop-down menus that allow the CAD user to hide or show a particular group of items, regardless of grouping method used. For example, most CAD systems have a readily accessible button to turn on and off the display of planes and annotations.

4) Change Dimensions Method: Controls the dimension value in each display schema, allowing the manipulation of component locations such as in a stowed and deployed mechanism.

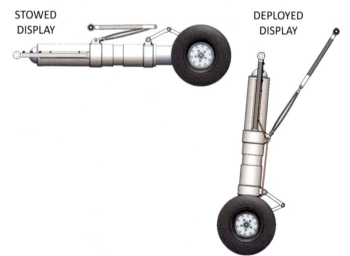

Example of stowed and deployed change dimension display schema method.

5) Transparency Method: Changing solid display of a component to assist viewing "inside" an assembly.

Example of transparency display schema method.

6) Large Assembly Visualization: Most CAD software has additional tools that assist in managing large assemblies beyond just suppressing fasteners. Large assembly visualization methods configure displayed models unique to the CAD native format in the effort to minimize assembly load time and reduce the overall assembly size (in bytes). Large assembly visualization capability within CAD creation software is relatively new (last 3-5 years) and not widely used. CAD users trained to leverage large assembly visualization techniques will improve the performance of their CAD assemblies.

BEST PRACTICE: Feature and Component Completeness

To this point, there are rules that apply to the WIP state and those that must be completed by pending signature review. Once released, the model morphs into the manufacturing authority model. It is necessary to have all features and components complete in order to be consumed by manufacturing.

The steps between the table top review (pending signatures) and manufacturing review are mostly tweaking modifications and not significant model reconfigurations; therefore being mindful of all the upfront rules necessary to complete by release when part design begins is always of benefit.

As the model evolves from the default planes into a completed part or assembly, the number of features or components included will evolve over time. For instance, I don't generally model all cutting tool radii until I have most of the part geometry set. However, the final part must contain the final radius for full definition, but while I work out the part geometry, I can omit those radii (that often fail or take time to regenerate in a feature tree) until I am ready to add notes and tolerances and submit for tabletop review (pending signature state).

ASME Y14.41 states, "The Design Model shall contain a complete geometric definition of the part.

A) Design models not fully modeled shall be identified as such.

B) Features that are not fully modeled shall be identified as such."

MIL-STD-31000A states, "When 3-dimensional models are required for a production level TDP, the models shall be complete, accurate, fully defined representation of the items and contain every feature the item being represented is intended to contain. All information necessary to adequately define the item shall be contained in, or associated with, the 3D digital model to include but not limited to dimensions, materials, tolerances, datums, drawing notes, revision data, etc."[55]

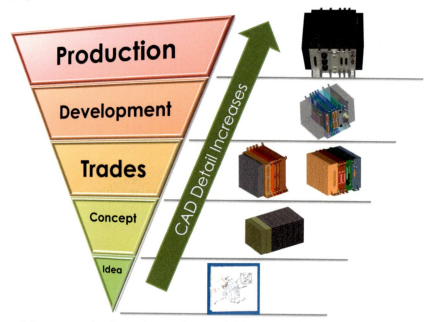

It is easy to mistake an incomplete model for a completely detailed model.

Over time, both parts and assemblies will go through an evolution as their designs are completed. To the naked eye, these models all look similar, but carry increasing (often concealed) detail. More explicit detail of part feature completeness and assembly feature completeness are covered in Chapters 6 and 7 respectively.

BEST PRACTICE: Modify, Rebuild and Check-In

 After any feature or component mate changes are made in a part, assembly, or drawing, you should rebuild or regenerate the CAD file. Iterate changes by making one change at a time and follow with a complete regeneration of the model. Giving your part a regular bath is good practice to minimize failure chaos implemented by feature or component change.

Deleting or Suppressing a feature or component are the most common examples of creating chaos in your part and generating errors because references get deleted or are temporarily removed.

TIP: GIVE YOUR PART A BATH

Check-in modified local files at least daily while making changes to that particular model. This process maintains a server-based backup and alerts other users who may be using your file that changes have been made. Regular daily check-ins will minimize confusion several weeks or months down the road because other users will be apprised of the changes ASAP.

BEST PRACTICE: Clean Data

 The CAD files promoted for Signature Release shall only contain features or components used in the model. Models shall not be a collection of all features ever created in a part. The PDM exists to capture iterative changes, so that the CAD files remain clean.

BEST PRACTICE: Design Modeling Intent

 Design Modeling Intent reflects how the model is intended to behave based on the desired design and the parameters (parametric values) you wish to adjust. For instance, you may need to adjust the number of fasteners in a hole pattern. Instead of modeling each hole as independent cuts, a pattern of holes that can be quickly modified may be preferred.

GOAL	BASE FEATURE	DATUMS	PLAN
Provide accurate, re-usable 3D CAD GEOMETRY	Anticipate what the base feature will look like	Anticipate where manufacturing datums will lie	Want to minimize do-overs

Basic instructions for planning any part or assembly model.

[6] Part Model Schema

Schema rules and best practices presented in this chapter only apply to individual part models. Assembly models are addressed in Chapter 7.

The part model schema is the recipe to make a solid model that represents an individual part.

 Part Model: A three-dimensional solid model representation of a single hardware item.

Each graphic shown represents a unique individual part model.

Envelope Model: A solid model representation of the outer envelope of an assembly that is represented as a single part.

For example, an electronics box is an assembly from a supplier; however, rarely do you need to display the internal electrical components. The outer envelope includes the box mount features, electrical connection interfaces, and protruding geometry, and are the only model geometry necessary to define the box's interface geometry.

FULL MODEL ASSEMBLY   ENVELOPE MODEL

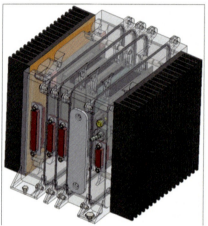 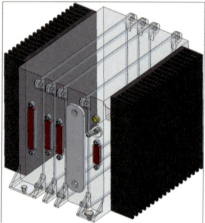

Example of the envelope model created from a full model assembly of an electronics box. The full model has a display schema called "ENVELOPE", facilitating repeat re-creation of the envelope model.

PART MODELING RULES

RULE: Number and Description

Number and description metadata elements give the part model its unique identification. Your organization will set the value and format desired for Number and Description.

 Number metadata value shall be entered at part model inception.

Description metadata value shall be entered by the time the part model is promoted to pending signature.

RULE: Material and Mass

Apply a material to every part model, referenced from the material library (recall Chapter 4). If a part is used as reference geometry only, then assign the part a material of "ZERO MASS".

ZERO MASS material density = 0.000000001 g/cm3 = or equivalent

Creating a ZERO MASS material in your material library allows all CAD designers to set the mass to an estimation of zero consistently throughout all parts.

Mass is not defined nor required in ASME Y14.41 to complete Model Based Definition (MBD); however, mass is a low-risk addition to each model, with high benefit for system analysis methods that need mass properties at the top assembly level. Without part models that contain a calculated mass, analysis of the top assembly for mass and mass properties is not possible.

Every part should have an assigned mass and shall be calculated by one of the following methods.

① Assign a material to the part model from the material library. Based on the density of the particular material chosen, part model mass, center of mass, and Moments of Inertia (MOI) will be calculated using the part model geometry.

② Assign mass, center of mass, and MOI explicitly to the part model using tools provided in the CAD software to manually enter the values for mass, center of mass, and MOI. Most CAD modeling tools will allow the designer to enter mass and then calculate volume of the part model to determine center of mass and MOI. This is a very handy method of mass property estimation when total mass is known but the part is an envelope model. While this method is not 100% accurate, it is a superior estimation technique to manual or spreadsheet calculations. Use caution when employing this method, as some CAD systems have generated errors when the parts with assigned mass properties roll-up into the top assembly.

RE-USE YOUR CAD: THE MODEL-BASED CAD HANDBOOK

Assign density to the model by using the desired mass and calculated model volume. The CAD software then calculates the center of mass and Moments of Inertia.

Density = mass/volume

RULE: Part Origin

The part origin is a point about which the primary features of the model are created. In most cases, model features are co-located at the default coordinates (x=0, y=0, z=0). However, there are exceptions to modeling features at the part origin, based on part function and size.

Small Parts (fasteners, brackets, connectors, switches, secondary structure)

Small parts that are added to an assembly to create a subassembly shall have their features modeled near or at the part origin.

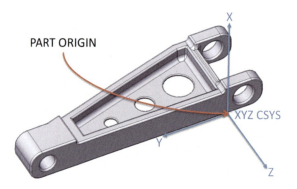

Linkage part model and coordinate system, where the features are modeled surrounding the part origin.

Large Parts (primary structure, vehicle envelopes)

It is often convenient for large components to be modeled at their final resting location with respect to the vehicle default coordinate system. Therefore, features in large models shall be modeled where they sit in space with respect to the vehicle coordinate system, rather than near or at the part origin, which is co-incident with the coordinate system origin (x=0, y=0, z=0).

The primary reason to model large parts in their vehicle space is that large part feature geometry may be referencing a vehicle envelope and intended for the development stage. Maintaining a direct connection to the vehicle envelope is often a method to achieve quick design iteration.

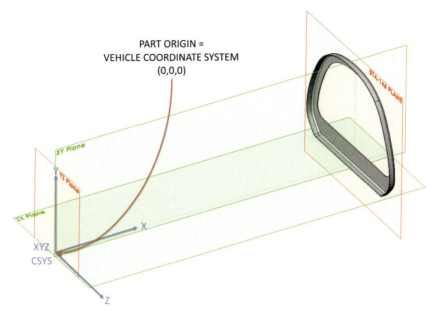

Example of a composite bulkhead model in vehicle space.

RULE: Feature Constraints

A model feature is a function used in CAD modeling software that defines a 3D shape, which, used in conjunction with other features, generates the complete solid model geometry that represents the part. Examples of features include but are not limited to: extruded shapes, holes, extruded cuts, revolved shapes, chamfers, and rounds.

For example, placing a sketch on a plane and identifying a vector that identifies the direction of pull for the extrusion creates a feature. It is necessary to un-ambiguously locate the sketch and pull vector relative to the default references and part origin. This is called constraining.

① Constrain model features to the highest reference possible, i.e. default planes, axis, and coordinate systems, rather than creating new reference geometry or placing the feature on top of preceding features. The reason for this is that the default reference frames will never be deleted and are therefore a reliable stable base on which to place your feature.

It is common when designing a part to move, change, or even delete features. If the last features created on a part reference the preceding feature, and the preceding feature is deleted, the last feature will most likely be affected, often leading to model rebuild errors.

Think of a Jenga® tower. The base feature (recall modeling intent from Chapter 4) becomes the stack of three blocks on the table and the next level of blocks, stacked 90 degrees to the lower level, represent the next features in the part model. Remaining features stack on the preceding features (Jenga® level) creating the full tower, the part model. You might get away with pulling out some of the middle blocks before the tower topples, but every piece you pull out supports the blocks above. The more you remove, the more risky your chances of the Jenga® tower toppling becomes.

 Only constrain feature references to preceding features if necessary and dictated by the modeling intent.

RULE: Sketch Constraints

 Most model features are created using 2D geometry (called a sketch) placed on a plane and extruded along a vector.

Constrain sketch geometry completely until the sketch is fully defined. Fully constrained sketches are the most robust. Most CAD software will alert you by changing the sketch color if the sketch is fully or under-constrained.

Partially Constrained Sketch – When a dimension is changed, the sketch shape is distorted.

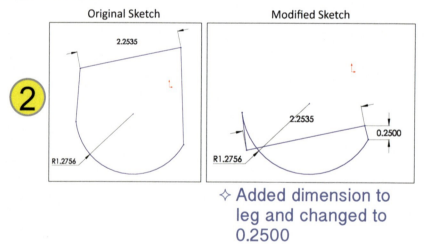

Example of a sketch that is not fully constrained.

Fully Constrained Sketch – When a dimension is changed, the sketch shape is preserved, and the CAD system alerts the designer that their modification is not legitimate.

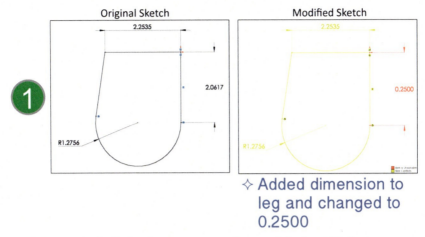

✦ Added dimension to leg and changed to 0.2500

Fully constrained sketches are more robust and the least likely to cause radical model failures or errors.

RULE: Feature Completeness

By the time the model is ready to be released for pending signature review, the model shall contain all features required to produce the part model with the following exceptions: threads, sharp corners, and broken edges. More details on feature exceptions are to follow.

Creating a complete set of features to define the model geometry unambiguously is required when generating digitally consumable documentation. It is critical to define the exact desired shape of the part after it is manufactured, as the geometry is no-longer "re-translated" through human interpretation. Mathematically accurate digital geometric representation is required to leverage emerging game changing manufacturing technologies, such as additive manufacturing.

Models that contain complete feature detail can be used downstream for:
1) **Form, fit, function, and interference checking.** *When complete feature geometry definition is missing from a model, a human must be in the loop to interpret the data. Human in the loop means missed opportunities to automate form, fit, function, and most especially interference checking.*
2) **Direct geometric transfer into manufacturing machines that are numerically controlled or are additive machines (3D printers).** *If a feature is not modeled in the design model, the resulting derived manufacturing model will also be missing that feature. Product detailing methods that rely on making modifications to 2D drawings, as*

the graphical modifications are not readable by any downstream manufacturing automation software, are not suitable for model-based methods.

Do complete models all have to be created by the design engineer? No. The advantage of a model-based environment is that your organization can leverage a concurrent engineering process. For example, "3D detailers" similar to 2D detailers could be utilized to create the product definition and remove detailing from the design engineers plate. 3D detailers can be responsible for final feature creation; the concept of having an engineer starting a model and having it completed by someone else is not hindered in a model-based environment. Rather it is enabled, as it allows transfer of source data between multiple individuals with the least amount of errors.

RULE: Radius & Chamfers

Model all fillets (rounds) and chamfers that are not broken edges.

To assist in faster model rebuild times and minimize feature rebuild failures, add radius and chamfer features as late as possible in the history tree. The Jenga® analogy again applies here, as radius and chamfers are small features that can wreak havoc in the feature tree, so it is best to leave them as the last block on the Jenga® tower.

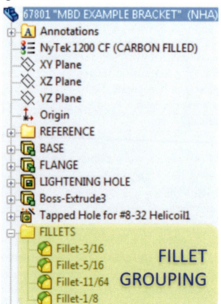

Part feature hierarchy showing the radius features left until the end.

RULE: Holes

There are two types of holes.

1) ***Standard sized hole*** that is re-used in many parts, such as a hole that a fastener goes through. This includes clearance, interference fit, and threaded holes.
2) ***Hole that is created as a cutout*** in the part and is non-standard or not intended to be re-usable in multiple parts. Often this type of hole is called a cutout, yet is still a cylinder shape.

There are two methods to create both hole types.

① Use the Standard Holes Library or Hole Wizard. The nomenclature of this feature will vary based on your CAD package.

② Sketch a hole with the sketch tool and cut the extruded sketch through the part.

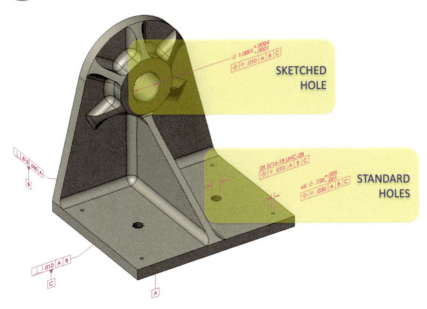

Example of Type 1 Hole (Standard) and Type 2 Hole (Sketched)

RULE: Threads

A modeled thread is created using a helical feature. Helical features are notoriously RAM intensive.

Do not model standard thread geometry.

Unless the threads are to be 3D printed directly from the model, there is no value added to modeling threads. In the future, computational speed and RAM may increase to facilitate thread geometry, but current technology limits the value and necessity of modeling threads. Though it seems trivial, this is vitally important, as one helical feature buried deep in a top assembly structure has the potential to eliminate the capability of a top assembly to load on even the fastest machines.

When reviewing the part model, how do I identify a threaded hole? Create an annotation attached to the hole feature with thread parameters. The combination of the hole wizard data and the 3D annotation tool should automatically generate the thread callout text. Restrain yourself from manually editing that annotation. Some CAD systems have the option to display a cosmetic thread (an image of a threaded feature or a hidden cylinder). Displaying cosmetic threads is optional and provides a human reviewer with a visual reminder that the hole is indeed a thread.

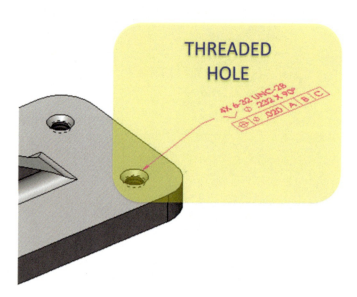

Threaded hole shown with a cosmetic feature, thread callout, and geometric position tolerance.

③ ***Thread exception: There is only one reason to model threads, which should be done so only as a last resort.*** Only when the threads are absolutely necessary, the following modeling rules shall be applied to the thread feature.

A helical (thread) feature must be:

- Placed in a group.

- Suppressed in ALL display schemas, even high fidelity display schemas. The thread feature shall never become part of the Next Higher Assembly (NHA) and is only of value for part manufacture.

- A special display schema must be created to un-suppress the threads only for the purpose of reviewing the part model and the thread feature.

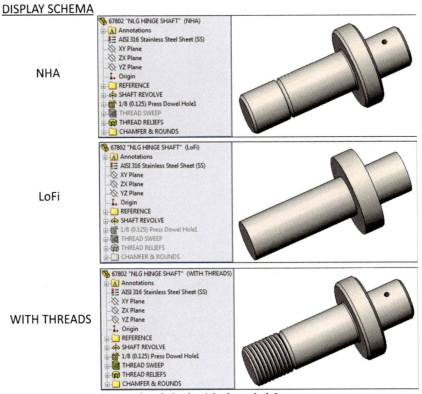

Example of shaft with threaded feature.

RULE: Webs and Flanges

Many products require webs and flanges used in a variety of functions. Some CAD systems have slick interfaces to create web and flange features. When selecting the references to create a web or flange, rely on base and/or preceding features. Apply webs and flange features as close to the end of the feature tree as possible. The reason again is because the features are stacked on preceding features, invoking the Jenga® analogy.

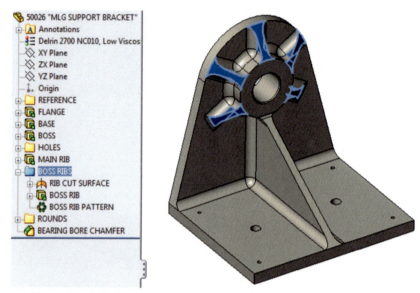

Example of ribs in a pattern and created late in the part feature tree.

RULE: Mirrored Features

Mirrored features are features that are reflected about a plane. They can be finicky depending on the CAD system. Because of this, apply the mirror function **as late as possible** in the feature history tree.

RULE: Mirrored Parts

A mirrored part is created when the entire completed part is reflected about a plane. Creating an associative link between the original part and the exact mirrored part is highly useful, as many top assemblies require a left and right hand part that are identically symmetric. When the original part is updated, the mirrored part updates as well.

RULES: Surfaces

Final part models (unless the part model is used FOR REFERENCE ONLY) shall not have surfaces as the final product. However, surfaces may be necessary for feature creation, but shall be thickened to create a solid model.

When practical, group like surfaces together in a group. Surface geometry not consumed in a feature shall be grouped and suppressed in the NHA display schema.

PART MODELING BEST PRACTICE

BEST PRACTICE: Groups

Expanding on the grouping best practice from Chapter 4, grouping for part models can be done by folders, layers, or groups, dependent on the CAD package. Sometimes CAD software can be configured to auto-filter certain feature types into specific folders, and some offer groupings as manual functions.

Folder Name	REQUIRED	What Goes in Here?
ANNOTATIONS	X	Tolerances, notes, and dimensions
REFERENCE	X	Default and additional planes, axis and coordinate systems
FILLETS		All rounds. Folder located at the end of the feature hierarchy.
HOLES		Clearance and threaded hole features, may include patterns of holes

Example of folders, required and optional, for a part model.

Group Together
- Annotations – REQUIRED OF ALL PARTS
- Reference geometry – REQUIRED OF ALL PARTS
- Similar features
- Patterns
- Combine features that are to be suppressed/un-suppressed and/or hidden/shown together

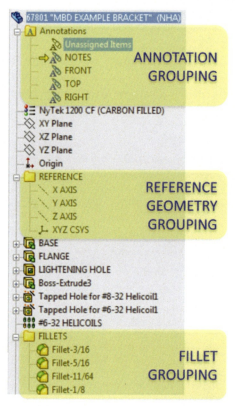

Example of part model hierarchy using grouping methods. Annotation and Reference folders are required; the Fillet folder is optional.

BEST PRACTICE: Feature Names

Instead of using the auto-generated name to describe the feature, rename features to make them descriptive of the feature. This aids the user later on when change requests drive a revision of the part model.

BEST PRACTICE: Display Schemas

Part models do not require any default display schemas, such as Low Fidelity (LoFi) configurations. However, some parts may require a LoFi configuration to be added because of its feature complexity. A compression spring model will need a LoFi configuration of a compression spring in order to represent the helical feature of the spring as a cylinder.

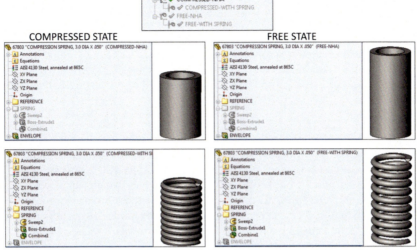

Compression spring part model showing compressed and free states in the same model. Each state has a derived display schema, which shows the full spring geometry.

BEST PRACTICE: Suppressed and Hidden Features

Recall the difference between suppressed and hidden features from Chapter 5 to assist in comprehending when suppressed and hidden features are permitted.

While a part is under development (in WIP state), it is reasonable and practical to have features modeled in the part that are not used in the final part model. However, by the time the part is released, all old or remnant features irrelevant to the final part design must be removed.

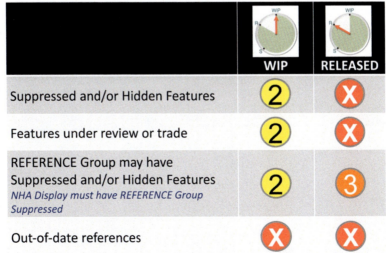

This table instructs users about when features can be suppressed and when a feature must be removed from the model.

BEST PRACTICE: Inter-part References

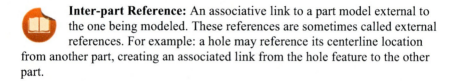

Inter-part Reference: An associative link to a part model external to the one being modeled. These references are sometimes called external references. For example: a hole may reference its centerline location from another part, creating an associated link from the hole feature to the other part.

Inter-part references are highly effective at ensuring that there is a single source geometry definition for an interface over several different part models. However, similar to circular references in an Excel spreadsheet, it is quite easy to get confused about what feature references what part, which can create a do-loop of pinging references when the part model is regenerated or loaded, causing long rebuild and load times. Therefore, the following guidelines should be considered before creating external references.

- Use inter-part references sparingly.
- Inter-part references easily get out of hand; be very specific and purposeful when creating these references.
- When copying and pasting a sketch, copy only the geometry and leave the references behind.
- Consider using a skeleton model to define highly referenced interface geometry.
- Lock external references when a part model is released.
- After creating inter-part references, review the inter-part references to check for relevance to design intent; if irrelevant, remove, freeze, or break them.

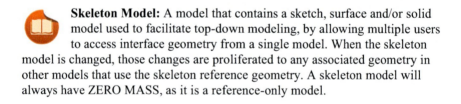

Skeleton Model: A model that contains a sketch, surface and/or solid model used to facilitate top-down modeling, by allowing multiple users to access interface geometry from a single model. When the skeleton model is changed, those changes are proliferated to any associated geometry in other models that use the skeleton reference geometry. A skeleton model will always have ZERO MASS, as it is a reference-only model.

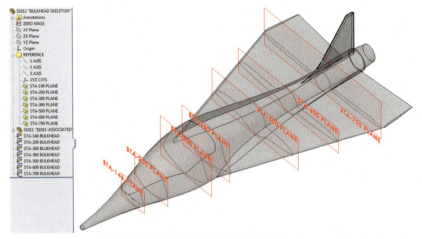

Skeleton model (51012-BULKHEAD SKELETON) shown with the Outer Mold Line (OML) surfaces. Each plane and sketch controls the cross section geometry for each bulkhead in this airplane.

BEST PRACTICE: Composites:

At this point, model-only definition of composites is a challenge. Use surface or solid models to define composite geometry and the drawing to identify layup plys, direction, and resin application. Generally, composite digital data definition will require a model + drawing data set for full product definition communication.

BEST PRACTICE: Direct Modeling

Direct Modeling: The capability to manipulate a 3D CAD surface or feature by drag and drop method, while maintaining solid geometry integrity.

Direct modeling is new and useful. However, don't mistake its uses for replacing the rules of best practice modeling. Instead, consider a hybrid method that combines feature-based modeling (and rules as described in this handbook) plus direct modeling methods. This is a very powerful combination. Some CAD systems take advantage of this hybrid approach.

When working in a hybrid environment, create features first, followed by direct modeling techniques.

PART TEMPLATE REQUIREMENTS

This chart identifies all the pieces that a part model template or "start part" should have based on the rules presented.

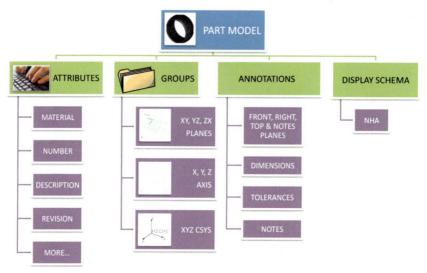

Required template components of a part model.

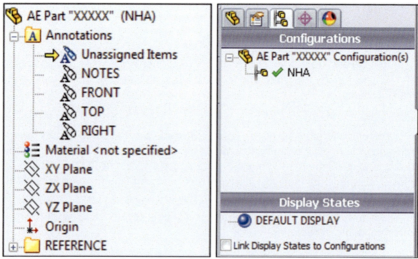

Example of SolidWorks part model template. Note that SolidWorks does not allow default planes to be grouped into the REFERENCE folder. Nevertheless, the intent of the rule is still met.

[7] Assembly Model Schema

Schema rules and best practices presented in this chapter apply only to assembly models. Part models are addressed in Chapter 6. Those rules and best practices common to both are included in Chapter 5.

The assembly model schema is the recipe for an assembly.

Assembly Modeling Philosophy

When assembly models are accurate and complete, assembly data sets can be used in a new and innovative fashion that enables downstream consumption of full product data.

Limits in both software and hardware technology have forced CAD users to adopt restrictions on complete assembly data sets. However, by carefully implementing part modeling best practices, it is now feasible to create complete and accurate assembly data sets.

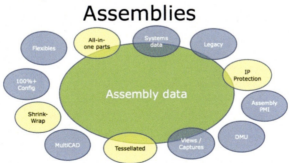

Accurate assembly models enable enhanced data usage possibilities. Graphic courtesy of Theorem Solutions, Inc.[56]

TERMS AND DEFINITIONS FOR ASSEMBLIES

Assembly: A number of parts or combinations thereof that are joined together to perform a specific function, and subject to disassembly without degradation of any of the parts (e.g., power shovel-front, fan assembly, audio-frequency amplifier). *NOTE: The distinction between an assembly and a subassembly is determined by individual application. An assembly in one instance may be a subassembly in another instance where it forms a portion of a higher assembly.*[57]

Example of an assembly of an electronics box.

Assembly Model: A collection of 3D solid models that represent the assembly and source data for the Parts List or Bill of Materials (BOM).

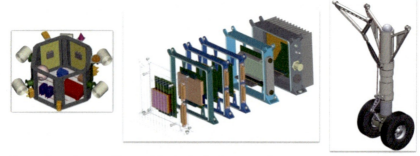

Examples of assembly models. Bolt, P-clamp and motor graphics courtesy of Cadenas Part Solutions.[58]

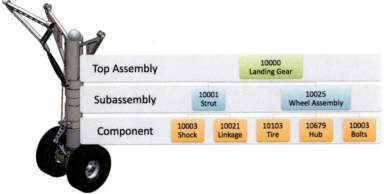

Example of an assembly model, representing nose landing gear from an airplane made of subassemblies and components.

Subassembly: Two or more parts that form a portion of an assembly or a unit, replaceable as a whole, but having a part or parts that are individually replaceable.[59]

Subassembly Model: A collection of part models that define a subset of a Next Higher Assembly (NHA) model.

Wheel subassembly model for nose landing gear.

Component: A part or subassembly model that is assembled into an assembly or subassembly model.

TIP:
Keep any assembly hierarchy to a maximum of 5 levels.

Each graphic represents a unique component of the nose landing gear assembly model.

Instance: Each instantiation of a component is called an *instance*. It may appear that the part model (component) is copied several times within a single assembly, yet the other representations are merely visualizations of the source data.

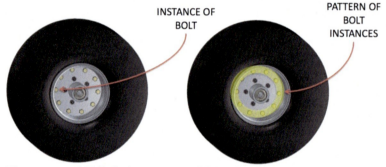

The pattern of 10 bolts is represented by 10 instances of the same bolt.

Assembly Constraint: An associative relationship created from a component to assembly references (planes, axis, or coordinate system), or to another component in the assembly for the purpose of defining the component location and orientation within the assembly.

Parts List: A tabulation of all parts and bulk materials used in the item(s), except those materials that support a process and are not retained, such as cleaning solvents and masking materials. NOTE: Other terms previously used to describe a parts list are a list of materials, bill of materials, stock list, and item list.[60]

Bulk Items: Those constituents of an assembly or part (such as oil, wax, solder, cement, ink, damping fluid, grease, flux, welding rod, twine, or chain) that satisfy one or more of the following criteria: the quantity required cannot readily be predetermined; the physical nature of the material is such that it is not adaptable to pictorial representation; the finished size is obtainable through use of such tools as shears, pliers, or knives, without further machining operation;

and the final configuration is such that it can be described in writing without the necessity of pictorial representation.[61]

Inseparable Assembly: A collection of two or more parts that are only disassembled by destruction of one or the other parts. These may be represented as a part or an assembly model.

Installation Assemblies: A collection of part and assembly models, assembled together for the sole purpose of describing the installation of a component or assembly. This assembly does not have a next higher assembly, nor is it typically the top assembly.

Often, an assembly model is created in order to describe the methods to install a unit (normally another assembly model) into a larger assembly (sometimes an entire vehicle). Designers generally call these assembly models installation assemblies. For instance, installing a passenger airplane seat into an airplane may require the CAD designer to create an assembly that contains the airplane seat, aircraft floor, and interior assembly, yet does not require the model details of the cockpit or wing assemblies.

The seat installation assembly does not have an NHA, but rather is created as a subset assembly of the entire vehicle to facilitate smaller assembly load times. Loading the entire airplane assembly is not generally necessary for airline seat installation procedures to be accurately depicted.

Installation assemblies evolved out of a drawing-based documentation process, where it was not necessary to visually show the entire vehicle to show a select few components of the top-down drawing tree. Nefarious, non-logical assemblies with no place to go confuse our human brains because they do not fall nicely onto a drawing tree structure, and tend to send zingers into a neatly organized product hierarchy. However, because the tools and technology may still limit complete top assembly loading onto a single computer, installation assemblies are often necessary evils.

ASSEMBLY MODELING RULES

RULE: Number and Description

Number and description metadata elements give an assembly model its unique identification. Your organization will set the value and format desired for Number and Description. Guidelines and tool considerations are covered in Chapter 5.

Number metadata value shall be entered at assembly model inception.

Description metadata value shall be entered by the time the assembly model is promoted to the status of pending signature.

RULE: Mass

If all parts have been assigned a material or density, as described in Chapter 6, then mass properties (mass, center of mass, and MOI) for the assembly will be correct. If not, then they will not.

Bulk items, such as: adhesive tape, wire wraps, and epoxy, have not traditionally been modeled, as they are not required in a drawing-based documentation system. Often, bulk items make a significant impact on overall assembly mass; therefore, creating a part model that represents that bulk item and its estimated shape is desirable.

Bulk item mass can be estimated in one of two ways in a model-based environment.

1) *Create a representative solid part model that estimates the bulk item shape and derive the mass from that estimated shape.* For example, creating a volume with the surface area shape matching that of the composite panel and the thickness equivalent to the nominal matte thickness can represent a matte adhesive on a composite panel. Give this part model the appropriate density value (as described in Chapter 6) to produce the desired mass. The resulting mass, center of mass and MOI calculated will be a reasonably close estimate.

2) *Create a part model with a general shape that approximates the volume of a collection of bulk items, apply a mass estimate and derive center of mass and MOI.* For example, to estimate a cable mass and its center of mass, create a swept volume that estimates the cable path. Combine the wire, cable wrap, and wire ties estimated density per linear inch together and apply the combined mass estimate to the swept cable volume envelope. The center of mass and MOI will be a reasonably close estimate and the mass will reflect your given mass estimate.

RULE: Assembly Origin

The assembly origin is similar to the part origin, where small and large assemblies shall determine how components are constrained relative to the assembly model origin.

1) **Small Assemblies** – Constrain components near the assembly model origin. Choose a base component that remains fixed in the assembly and "ground" it to the model origin. Small assembly examples are: mechanisms, electronics boxes, and replaceable units.

2) **Large Assemblies** – Constrain components relative to the vehicle coordinate system as dictated by their component location within the

larger assembly. A large assembly is one that contains components that sprawls over a large portion of the top assembly. In this case, consider the model origin as "ground" for the large assembly. Large assembly examples are: primary structure, fuel systems, cabling systems.

RULE: Component Constraints

When a part or subassembly is inserted into an assembly, it becomes a component of the assembly. That component shall be unambiguously constrained within the assembly.

The first or base component of an assembly shall be constrained to the default planes, axis and/or coordinate system.

Subsequent components shall be constrained as follows:

 Constrain the component to parent assembly default planes, axis, or coordinate system.

 Constrain the component to a preceding component or feature as required by the design intent. For example, a bolt in a clearance hole justifies a pin in the hole or concentric mate between the part with the clearance hole and the bolt.

TIPS:
a) Avoid creating a chain of mates to preceding parts. Remember the Jenga® tower? – This is the same concept.
b) Concentric or Center Alignments – Align to the originating center feature.
c) Plane Mates and Alignments – Align to parent planes.

RULE: Assembly Model Completeness

The level of completeness of your assembly model determines how well others will be able to collaborate with and integrate your assembly model into their system models.

As the design development process progresses, so should the completeness of the assembly model. This includes not only parts and geometry accuracy, but also constraint relationships and adherence to standards.

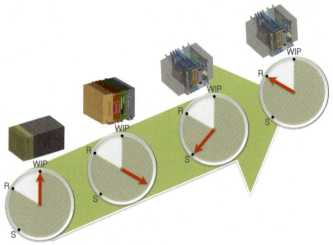

How complete an assembly model is will grow over time. However, discerning assembly model completeness requires examining the assembly model in 3D. 3D reviewing may be uncomfortable and disorienting to users. Allow for adequate orientation for your users to become accustomed to this practice.

   An assembly model is complete when:

- All parts and subassemblies required to build the assembly are assembled as components in the assembly model.
- All components are unambiguously constrained.
- All fasteners are included and constrained. *Note that for cylindrically symmetric parts such as a screw, it is not necessary to constrain the rotation of the screw about its axis. Only two constraints are required for the screw to constrain it in the assembly reliably: mating face to underside of bolt head and concentric or axis mate.*

Why is it important to include a component in an assembly model that represents every item in an assembly?

1) **Parts list source:** *A complete assembly model serves as the data source for the parts list.*
2) **Accurate assembly model mass**
3) **Accurate form, fit and function checking at the NHA**
4) **Complete assembly product definition**

RULE: Parts List

A part list only applies to an assembly model.

What the industry typically calls a BOM and puts on a drawing is officially called an *associated list*. Some companies prefer the BOM to be physically on the drawing or displayed in the drawing; this is called an *integral part list*.

However, those that have moved to using a PLM system as a product release and documentation tool auto-generate the BOM from part metadata, removing the BOM from the drawing or model and maintaining it in a separate part list. A best practice is to maintain an associative connection between the parts list and the assembly it describes.

The parts list is to be delivered with the assembly model as part of the TDP.

	Mechanical Bill of Materials (MBOM)	**Engineering Bill of Materials (EBOM)**
COMPONENTS (Parts and Subassemblies)	X	X
BULK ITEMS	May have a model representation	X
CONSUMABLES		X
SPECIFICATIONS		X

Flowing assembly model components into the Mechanical BOM and then to the Engineering BOM provides significant advantages to reduce data entry errors and to reduce duplicating work.

Parts List Documentation Possibilities

 Integral Parts List --- There are a variety of options for using integral parts lists with an assembly, and the options vary based on data set deliverable type.

The following are examples of integral parts list implementation options.
- Model + Drawing data set type: Table on the drawing
- Model-Only data set type:
    - Table on the assembly model
    - Table in a 3D PDF with associative connections to assembly model (represented in the 3D PDF)

Separate Parts List --- A separate parts list is one where the list is represented in a table and stored separate from the assembly. It may or may not source from the assembly model or have an associative relationship.

The following are examples of separate parts list implementation options.
- Excel spreadsheet with find numbers
- PDM and/or PLM database view-only (not associated to the data set) with find numbers

Example of an integral parts list displayed in native software format for a model-only product definition for landing gear.

RULE: Assembly Sketches & Constraints

Assembly sketches and constraints shall be unambiguously defined.

Most CAD systems allow sketch geometry to be created at the assembly level. Assembly level sketches follow the same rules for constraining as they do when generating a part feature. A fully defined sketch is the most robust sketch.

Refer to Chapter 6 for further details of constrained versus unconstrained sketches.

RULE: Assembly Feature Completeness

Assembly features are not particularly common, but when used can be a very useful tool. Listed are a few examples that would require an assembly feature.

- Holes cut through several parts at once, such as when two mating parts are match drilled.

- Cutouts made through several parts at once, such as when routing a composite panel.

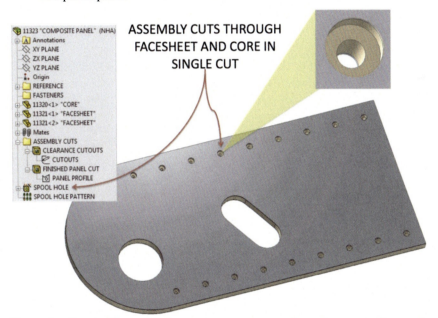

Example of assembly cut used to create a representation of a composite panel.

When assembly features are created, they require a sketch. That sketch follows the same sketch constraint requirements as described for part models in Chapter 6.

Because it is important to minimize inter-part relationships and rebuild assembly times, careful attention must be paid to where and how the assembly sketch is constrained.

 Constrain to highest reference possible, i.e. default planes, axis, or coordinate system.

 Only constrain to preceding features or components if necessary, and if necessary to meet the design intent. The more your features and components "touch" other components, the larger the assembly file grows, and the more complex the inter-related spider web of data grows.

TIP: A carefully laid pattern of yarn can easily be untangled, but a randomly knotted ball of yarn, will be virtually impossible to unravel.

RULE: Assembly Patterns

A pattern of components shall be used in preference over inserting multiple instances of a single component.

A properly setup pattern of components "points to" geometry from another component in the assembly. For instance, a pattern can be created of a bolt component that is inserted into a hole pattern in a bracket. By choosing to use the pattern in the part, to which the bolt mates, the parent pattern in the bracket controls the number of bolts at the assembly automatically.

Using component patterns also assists the CAD designer in keeping parts consistent in an assembly by ensuring that multiple models that represent the same part are not inserted into the assembly on accident. Because the pattern involves a single component with multiple instances, it will populate the parts list (or BOM) accurately. In some CAD systems, multiple instances of the same component may show up as separate line items in the parts list.

- Use patterns of components rather than inserting multiple instances.
- Patterns assist in insuring instances are identical, rather than a slightly different model.

ASSEMBLY MODELING BEST PRACTICE
BEST PRACTICE: Groups

Expanding on the best practice presented in Chapter 4, that of creating logical groups, assembly models also gain great benefit from groups of common components and features. Some CAD software can be configured to auto-filter certain component types – fasteners for example.

	Folder Name	REQUIRED	What Goes in Here?
📁 ANNOTATIONS	ANNOTATIONS	X	Tolerances, notes, dimensions, parts list
📁 REFERENCE	REFERENCE	X	Default planes, axis, coordinate systems, additional reference planes, axis, coordinate systems, sketches, reference components, skeleton models
📁 FASTENERS	FASTENERS	X	All fasteners: bolts, screws, nuts, washers, rivets, cotter pins, etc.
📁 COMPONENTS	COMPONENTS		Small components such as bearings, p-clamps and switches
📁 ASSY CUTS	ASSEMBLY CUTS		Features created at the assembly level that affect multiple components.

Examples of folders, required and optional, to facilitate groups for assembly models.

Group Together:

- Annotations – REQUIRED OF ALL ASSEMBLIES
- Reference geometry and components – REQUIRED OF ALL ASSEMBLIES
- Similar components
- Assembly features and patterns
- Combine components that are to be suppressed/un-suppressed and/or hidden/shown together. Examples of component or assembly features that might be grouped together: fasteners, shims, cabling, datums, and assembly cuts.

TIP: Folders and Groups are the preferred grouping method, rather than using layers.

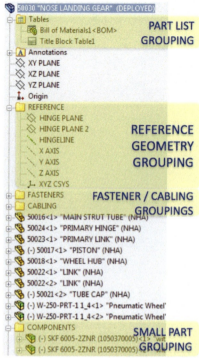

Example of assembly model hierarchy with groups represented by folders.

BEST PRACTICE: Display Schemas

Assembly models require two default display suppress/un-suppress schemas:

1) High Fidelity Display
- Default configuration or representation
- Generally feeds into the NHA
- Stowed
- All components in BOM visible and un-suppressed

2) Low Fidelity
- Small fasteners, components, and reference folder are suppressed
- This display schema is referenced by the NHA to achieve reduced load times

Assembly models may require additional display schemas, such as: stowed, deployed, or tooling configurations.

TIP: Some CAD software allows display schemas to derive from a parent schema. This is often handy, but not required.

Example of HiFi and LoFi display schemas in the deployed schema.

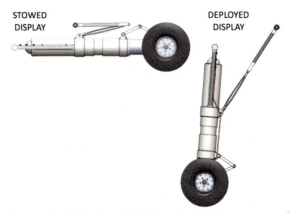

Example of stowed and deployed display schemas in one assembly model.

BEST PRACTICE: Suppressed and Hidden Features

Recall the difference between suppressed and hidden components from Chapter 5. Hidden components are still held in memory, and only removed from view on the screen. Hide and show display schemas facilitate a quick method to turn "on" and "off" components, as the part or assembly model is not removed from RAM.

Suppressed components are purged from RAM and improve load times of the assembly. However, it will take time to "turn back on" a suppressed component.

While the assembly is under development (in WIP state), it is a practical practice to have components assembled into the assembly, which are not in the final parts list. However, by the time the part is released, all old or remnant components, irrelevant to the final assembly model, must be removed.

Become a pack rat some other way than hoarding components in your assemblies that are no longer relevant. Create a working assembly that serves as your sandbox, but these sandbox assemblies should never be shared with others – EVER!

If sharing sandbox models is an acceptable practice, it creates a circular reference nightmare. Do not subject other users to your hoarding habits.

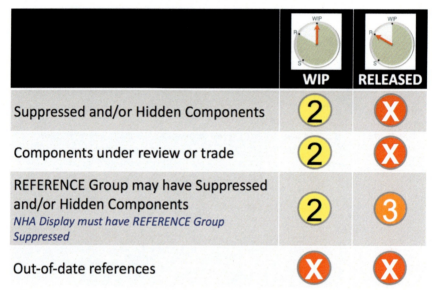

This table instructs users on when components can be suppressed and when the components must be removed from the model.

BEST PRACTICE: Assembly Relationships

Assembly references include constraints created from a component to the assembly planes, axis or coordinate system, and also any relationship from one component to another component within that assembly or another assembly.

Recall that an inter-part reference is an associative link between two or more parts. A single assembly may contain multiple inter-part references, some of which are component to component relationships and some that are component to assembly default references. For instance, an assembly hole feature created to show a match drilling operation at the assembly level associates the two parts (component to component), which are to be drilled together during manufacturing.

Minimize assembly references as much as allowed by the design intent.

When an assembly reference is created, a relationship link is generated from one part to another. A well set-up PDM will assist to easily manage these relationships.

A common and extremely useful technique is to reference geometry from another part that remains associative. Thus the part created using the geometry from another part (the parent part), updates when the parent part modifies.

Beware that this technique creates inter-part relationships and will increase load times of the assembly when those relationships are broken or difficult to reference. For example, if the referenced part has been renamed or relocated, the system may need to search and find it before loading the reference information, delaying the load time of the assembly.

Different data management (PDM) systems handle associative relationships better than others. If your organization is a system integrator of large assemblies, determine your CAD/PDM software's limitations and prepare a plan to integrate subassemblies into top assemblies.

> *TIP: Minimize load times by managing references within the assembly. Follow inter-part reference best practice laid out in Chapter 6.*

The trouble with most assembly load times is not the geometry or individual component file size; instead, it is the inter-part relationships creating many pings to other files when opening or re-building or re-generating.

The following are methods for minimizing load times by managing references within the assembly:

- Constrain components to parent assembly default planes, coordinate systems, and axis
- Constrain to sibling components only when necessary
- Keep geometry references to only components within the assembly
- Keep geometry references to skeleton model geometry strictly controlled

BEST PRACTICE: Assembly Setup and Maintenance

Set up assemblies to the best of your knowledge. Refer to the design intent workflow presented in Chapter 5. Be thoughtful about which component is an appropriate "ground" for the assembly, and build from there.

A good system engineering team will have created a drawing tree for the product developments on which you are working. Use this drawing tree as a straw man to organize your assembly hierarchy. You may find that the drawing tree or the assembly models being developed begin to diverge once in development. I recommend re-aligning the assembly model with the drawing tree on regular intervals.

Like your checkbook, it is prudent to balance your assembly at regular intervals. When I leave my checkbook unbalanced for three months, it is very confusing and time consuming to reconcile. Rather, if I reconcile the checkbook on regular and frequent intervals (every 2-4 weeks), then reconciliation is relatively painless and is over with quickly. Additionally it is easier to remember where and what I spent my money on after only a few weeks, rather than after a few months.

Practice regular maintenance of assembly constraints and inter-part relationships, as they are a complex network of related data. Keep components updated to the latest version within the assembly and update the assembly version (check it in) on regular intervals. This practice protects users from losing data by creating robust backups as well as pushing updated product designs out to collaborative users.

BEST PRACTICE: Large Assemblies

Use large assembly and visualization tools to maximize efficiency for assembly modification and review.

Large assembly tools are lightweight suppression modules native to the particular CAD software. Visualization tools are those used for design review and delivery such as: eDrawings, CREO View Express, AutoVue, 3D PDF, and Teamcenter Viewer.

This tool facilitates display schema method #6: Large assembly visualization, described in Chapter 5.

ASSEMBLY TEMPLATE REQUIREMENTS

This chart identifies all the pieces that an assembly model template or "start assembly" should include based on the rules presented.

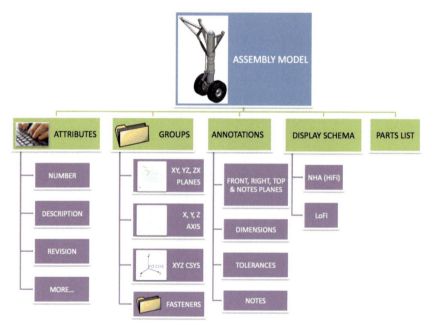

Required template components of an assembly model.

RE-USE YOUR CAD: THE MODEL-BASED CAD HANDBOOK

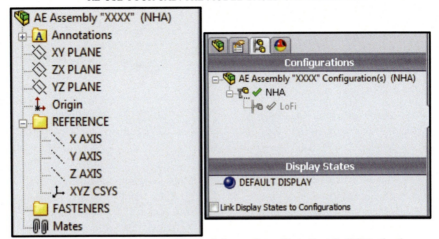

Example of SolidWorks assembly model template. Note that SolidWorks does not allow default planes to be grouped into the REFERENCE folder. Nevertheless, the intent of the rule is still met.

Epilogue

Almost every rule has an exception, but when we make exceptions to the rule, we shoot ourselves in the foot. A model-based documentation strategy has un-quantifiable benefits. When a model-based ecosystem is your rule, then your organization will start to reap MBE benefits.

The good news about MBE is that MBE does not eliminate drawings; rather it enables accurate re-use of model data to flow more readily and accurately into 2D drawings. This mindset provides a method to remove 2D drawing creation from the time-consuming engineering release process. If desired, 2D drawings can become a permanent archive source.

Let's say you are uneasy about documentation in digital form only. Then, generate 2D drawings from your model's 3D annotations and geometry, but do it outside of the release-cycle loop, and even better, automate that archival process.

The medium of communicating the product definition isn't the time-consuming part. Product definition must happen regardless of the delivery medium (drawing or model) --- this is the part that is time consuming.

I'm generally asked if moving to a model-based documentation strategy is a slam-dunk today. No it is not, but it is a slam-dunk in 5 years. The big advantage of implementing a model-based ecosystem now, is to be a leader and not wait until the time you are forced to be a follower.

I'm also asked if it is easy to mess up MBE implementation. It certainly is! Experience and complete comprehension of the technology pros and cons is vital to success.

It is easy to want a model-based architecture to accommodate as many disciplines as possible (i.e. systems level requirements and analysis models), which is a fantastic goal. However, I implore you to carefully hammer out the details required in model-based implementation for product definition release first, long before your "go-live" date, and before integrating other discipline requirements into the architecture.

In contrast, avoid analysis paralysis. I recommend a healthy monitoring and tweaking of the entire model-based ecosystem, tools and process for two years after the "go-live" date, as these first years of implementation will be the most important to nurture your culture into adopting new standards and processes.

Finally, I ask you to consider that model-based documentation become the rule and drawing-based documentation become the exception.

If a picture is worth a thousand words, then a 3D solid model is worth a trillion!

About the Author

The CAD Agnostic

Jennifer B. Herron is the owner of Action Engineering, a company that specializes in the promotion, process development, and standardization of 3D CAD Model-Based Design. Her career has been spent creating and building complex hardware systems for the aerospace and defense industry, her experience runs the gamut from flight hardware mechanisms to spacecraft layout and configuration. She is an expert in multiple CAD packages, which she uses along with her practical design experience to hone standards and processes that optimize the ROI of all CAD systems. In addition to her involvement developing many flight satellite systems, Jennifer has designed military robot platforms, holds a patent for a snake propulsion mechanism, has a Bachelor of Science in Mechanical Engineering, and a Masters of Science in Computer Engineering.

End Notes

[1] NIST Technical Note 1753 (August 2012), 1.0

[2] http://model-based-enterprise.org/Starting-Model-Based-Enterprise/default.aspx

[3] http://model-based-enterprise.org/Starting-Model-Based-Enterprise/default.aspx

[4] NIST Technical Note 1753 (August 2012)

[5] NIST Technical Note 1753 (August 2012)

[6] NIST Technical Note 1753 (August 2012)

[7] PMBOK, Chapter 8, Cost of Quality

[8] Robert Green, Expert CAD Management—The Complete Guide (page 107)

[9] Daniel Lipkowitz, The LEGO® Book, (page 37)

[10] Robert Green, Expert CAD Management—The Complete Guide (page 95)

[11] Robert Green, Expert CAD Management—The Complete Guide (page 51)

[12] Robert Green, Expert CAD Management—The Complete Guide (page 53)

[13] Sheryl Sandberg, Lean In

[14] Robert Green, Expert CAD Management—The Complete Guide (page 22, Fig 2.3)

[15] 2012 Supplier Feedback on the 3D Technical Data Package, http://model-based-enterprise.org/2012-assessment.aspx

[16] 2012 Supplier Feedback on the 3D Technical Data Package, http://model-based-enterprise.org/2012-assessment.aspx

[17] ASME Y14.41 (R2012), 3.21

[18] NAS 9300-007 (30-May-2008), 3.1

[19] MIL-STD-31000A (26 February 2013), Appendix B.3

[20] ASME Y14.41 (R2012), 3.11

[21] ASME Y14.41 (R2012), 3.6

[22] DEDMWG-MBE (2010-Aug)

[23] Advanced Dimensional Management, www.advanceddimensionalmanagement.com

[24] Advanced Dimensional Management, www.advanceddimensionalmanagement.com

[25] Advanced Dimensional Management, www.advanceddimensionalmanagement.com

[26] DEDMWG-MBE (2010-Aug)
[27] ASME Y14.41 (R2012), 3.24
[28] MIL-STD-31000A (26 February 2013), 3.1.37
[29] Mil-Std-31000A (26 February 2013), 5.3.3
[30] Mil-Std-31000A (26 February 2013), Appendix B.5
[31] Mil-Std-31000A (26 February 2013), Appendix B.5.1
[32] Mil-Std-31000A (26 February 2013), Appendix B.5.2
[33] Mil-Std-31000A (26 February 2013), Appendix B.5.3
[34] Mil-Std-31000A (26 February 2013), Figure 2
[35] John G. Falcioni, Editor-in-Chief, Mechanical Engineering, The Magazine of ASME. June 2013, No. 6, 135.
[36] NAS 9300-007 (30-May-2008): 3.1
[37] NAS 9300-007 (30-May-2008): 3.1
[38] MIL-STD-31000A (26 February 2013): C.2
[39] www.merriam-webster.com
[40] http://www.storagereview.com/ssd_vs_hdd
[41] ASTM F2792 (2012, Rev A)
[42] ASME Y14.41 (R2012), 3.21
[43] ASME Y14.41 (R2012), 3.12
[44] ASME Y14.41 (R2012), 3.22
[45] ASME Y14.41 (R2012), 3.31
[46] Advanced Dimensional Management, www.advanceddimensionalmanagement.com
[47] MIL-STD-31000A (26 February 2013), Appendix B.3
[48] ASTM F2792 (2012, Rev A)
[49] http://www.planetarysystemscorp.com
[50] MIL-STD-31000A (26 February 2013), 4.14
[51] Advanced Dimensional Management 2008
[52] ASME Y14.41 (R2012), 9.1.1
[53] ASME Y14.41 (R2012), 9.1
[54] Advanced Dimensional Management, www.advanceddimensionalmanagement.com
[55] Mil-Std-31000A (26 February 2013), 5.2
[56] www.theorem.com
[57] ASME Y14.100 (2004): Sec. 3
[58] www.partsolutions.com
[59] ASME Y14.100 (2004): Sec. 3
[60] ASME Y14.34 (2008), 3.20
[61] ASME Y14.100 (2004): Sec. 3

Made in the USA
San Bernardino, CA
12 January 2018